Springer Series on

Atomic, Optical, and Plasma Physics 77

For further volumes:
http://www.springer.com/series/411

Springer Series on

Atomic, Optical, and Plasma Physics

The Springer Series on Atomic, Optical, and Plasma Physics covers in a comprehensive manner theory and experiment in the entire field of atoms and molecules and their interaction with electromagnetic radiation. Books in the series provide a rich source of new ideas and techniques with wide applications in fields such as chemistry, materials science, astrophysics, surface science, plasma technology, advanced optics, aeronomy, and engineering. Laser physics is a particular connecting theme that has provided much of the continuing impetus for new developments in the field. The purpose of the series is to cover the gap between standard undergraduate text books and there search literature with emphasis on the fundamental ideas, methods, techniques, and results in the field.

Please view available titles in *Springer Series on Atomic, Optical, and Plasma Physics* on series home page http://www.springer.com/series/411

François Frémont

Young-Type Interferences with Electrons

Basics and Theoretical Challenges in Molecular Collision Systems

With 206 Figures

Springer

François Frémont
University of Caen
Caen Cedex
France

Springer Series on Atomic, Optical, and Plasma Physics

ISSN 1615-5653
ISBN 978-3-642-38478-3 ISBN 978-3-642-38479-0 (eBook)
DOI 10.1007/978-3-642-38479-0
Springer Heidelberg New York Dordrecht London

Library of Congress Control Number: 2013944660

Printed on acid-free paper

Springer is part of Springer Science+Business Media (www.springer.com)

Preface

We know that light has fascinated philosophers, physicists, theologians, and engineers for a long time. The nature of light, especially, has been the subject of many controversies and discussed in detail since the seventeenth century. This is essentially due to the fact that light in itself is not visible. From simple considerations, Isaac Newton put forward the idea that light was composed of little balls. This truly innovative idea was quickly countered with the emergence of the wave theory of light. For two centuries, this theory was proven in all areas of wave frequencies. However, at the beginning of the twentieth century, this ingrained idea was challenged through the decisive contribution of Albert Einstein, who showed that the photoelectric effect could be easily explained when light was considered to be composed of small particles, or "quanta."

Then, in 1924, Louis de Broglie formulated his well-known hypothesis that waves could be associated with any massive particle. Thus was born the concept of wave-particle duality. This duality particularly manifests itself in the interference phenomenon, first discovered for light by Thomas Young in 1801. Since the de Broglie hypothesis, many interference experiments using massive particles, such as electrons, protons, and atoms, have confirmed both the wave and particle theories.

Despite experimental and theoretical efforts, the nature of light and, consequently, the nature of massive particles remain mysterious. The very issue of duality means that we do not know how to define these particles. With the advent of new technologies, we could imagine that the nature of light would be revealed. But despite these efforts, the question remains. The only thing we can characterize is the behavior of these particles or waves, depending on their action on the interacting medium. Both aspects are nevertheless mixed: when wave behavior is revealed, the particle aspect is not lost, and vice versa.

Centuries of research, heated discussions, and controversies force us to make a detailed analysis of the past and the present situation. To the best of our knowledge, no book has dealt with the interference phenomenon, including light and massive particles. Consequently, analogies between photon interferences and massive particle interferences are rare. However, over the past decade, new approaches have been developed through the detailed analysis of interference figures produced by electrons emitted during fast or slow ion collisions with molecules.

To explain these analogies, this book is divided into five parts. The first and second chapters are devoted to interferences with light and massive particles, respectively. In the third chapter, we focus our attention on electron interference experiments using macroscopic and nanoscopic interferometers, which have been carried out since 1925. Particular attention is paid to what are referred to as the Young-type double-slit experiments that were performed in the early 2000s. We shall see that this designation, which refers to the famous experiment by Young in 1807, is in fact not accurate.

Chapter 4 describes a detailed analysis of a single-electron Young-type double-slit experiment. This experiment, based on low-energy He^{2+} + H_2 collisions, was theoretically described in 2004. During the collision, the He^{2+} ion targets captures both target electrons onto doubly excited states. After the collision, one electron is emitted from the projectile due to the Auger effect and scatters on both of the protons acting as the double slit. The ways to obtain a single-electron condition are discussed. The angular distributions of scattered electrons, as well as their energy profiles, are analyzed. A simple model, referred to as the Path-Interference model, based on the possible trajectories taken by the electron to reach the detector, is used to give a qualitative description of the angular distributions.

Due to the limitation of the previous model, which assumes that the electron is emitted at a given distance from the slits, a more refined analysis is made in Chap. 5 using the Final-State Interaction model. This model is based on a quantum description of the Auger effect. Using a Continuum Distorted Wave approximation, the energy profiles of the emitted electrons, as well as their angular distributions, are calculated. We shall see that, contrary to predictions, this model is unable to explain the interference pattern observed experimentally.

Finally, an attempt is made to describe the experimental interferences using a semi-classical approach. The orientation of the molecule, the time at which the electron is emitted, and the orientation of the electron velocity are randomly chosen. The Hamilton equations are solved numerically and electron trajectories are calculated. Then, at a fixed detection angle, the wave aspect of the electron is taken into account to calculate the phase shift induced by the delay in the trajectories, and the angular distribution of the emitted electrons is deduced. We shall see that, surprisingly, this model is promising, challenging the way we view the electron and the associated interference.

Caen Cedex, France François Frémont

Acknowledgements

I am indebted to Raul O. Barrachina, a theoretical physicist interested in atomic collision, historian of science, and passionate about everything related to humans, for motivating me to write this book. He succeeded, patiently and carefully, in transferring his thirst for truth to me. The time he took for it is priceless.

I also thank Jean-Yves Chesnel, without whose interference experiment with a single electron source could not have been possible; Maxime Vabre, who was at that time a second-year university student; and Sylvain Girard, Hervé Giles, Philippe Leprince and Florent Porée, who lead the experience of photon interference with Lloyd's mirror.

Many first- to fifth-year students worked hard on the semi-classical model: Guillaume Oliviero, Lucie Bottey, Florine Minerbe, Valentin Pestel and Méghann Philippe. I am proud to have worked with them and see their unwavering enthusiasm.

Finally, my thanks to Helen McCombie, from Brest University, who proofread the book; Springer Verlag publishers who accepted this project; and Ms. Dhivya Geno, project manager from Manapakkam, India, who has shaped the book.

Contents

1 Photon Interferences: History and Fundamental Aspects 1
1.1 The History of Light 1
1.1.1 Before Newton 1
1.1.2 Newton and Huygens: Two Opposite Approaches 3
1.1.3 Young: The Discovery and Formulation of Interferences 5
1.1.4 Michelson and Morley: A Genius and Decisive Experiment 7
1.1.5 The Quanta Revolution and Corpuscular Vision 8
1.1.6 Two Crucial Experiments 9
1.2 Characteristics of Interferences 11
1.2.1 Conditions for Interferences 11
1.2.2 Coherence in Time and in Space 13
1.2.3 Contrast and Visibility of Fringes 13
References 17

2 Interferences with Massive Particles 19
2.1 The De Broglie Wave-Particle Duality 19
2.1.1 The Concept of Wave-Particle Duality 19
2.1.2 The First Evidence for Wave Behavior of Massive Particles 20
2.2 Interferences with Massive Particles 20
2.2.1 Interferences with Neutrons 20
2.2.2 Interferences with Atoms 23
2.2.3 Interferences with Small Molecules 28
2.2.4 Interferences with Large Molecules and Clusters 30
2.2.5 Interferences with Bose-Einstein Condensates 32
References 33

3 Electron Interferences Using Macroscopic and Nanoscopic Interferometers . . . 35
3.1 Electron Interferences with Macroscopic Interferometers . . . 35
3.1.1 Interferences Using an Electron Beam: The Young Experiment Revisited . . . 35
3.1.2 Feynman Thought Experiment . . . 36
3.1.3 Interferences Using an Attenuated Electron Beam . . . 37
3.2 Electron Interferences with Nanoscopic Interferometers . . . 41
3.2.1 Fundamental Interest of the Approach . . . 41
3.2.2 Processes in Atomic Collisions . . . 42
3.2.3 Approximations in the Collision and Post-collision Calculations . . . 45
3.3 The Role of Electrons in Ion Interferences . . . 46
3.3.1 Elastic Process in Asymmetric and Symmetric Collisions . . . 46
3.3.2 Resonant Electron Capture in Symmetric Collisions . . . 49
3.3.3 Electron Capture in Asymmetric Collisions . . . 53
3.4 Electron Emission in Atomic Collisions: Young 1801 . . . 57
3.4.1 Photoionization of Molecules . . . 57
3.4.2 Ionization of H_2 by Ion Impact . . . 64
3.4.3 Electron Capture . . . 83
3.4.4 Ionization of H_2 by Electron Impact . . . 85
3.5 Electron Emission in Atomic Collisions: Young 1804 . . . 91
3.5.1 Principle of the Experiment . . . 91
3.5.2 Interpretation of the Spectra . . . 92
3.6 Other Sources of Interferences . . . 94
3.6.1 Interferences Between Direct Ionization and Autoionization . . . 94
3.6.2 Interferences Between Auger Lines . . . 96
References . . . 98

4 Young-Type Electron Interferences Using Single Electron Sources . . . 101
4.1 Principle of a Single-Electron Experiment . . . 101
4.1.1 Analogy with Photon Experiments . . . 101
4.1.2 An ideal Single-Electron Experiment . . . 101
4.1.3 Visibility of the Expected Fringes . . . 103
4.1.4 Limitations of the Contrast . . . 104
4.1.5 Coherence Time and Coherence Length . . . 106
4.2 Realization of the Experiment . . . 106
4.2.1 One and Two Event Probabilities . . . 106
4.2.2 The Experimental Setup . . . 108
4.2.3 Spectral Analysis for 30 keV He^{2+} + H_2 Collisions . . . 109
4.2.4 Evolution of Oscillations with Interference Parameters . . . 112

4.3 The Origin of the Oscillations in Electron Interferences 114
4.4 Analogy with the Photon Interference Experiment 117
4.4.1 Relevance of the Analogy 117
4.4.2 The Young Double-Slit Experiment with Photons Revisited 118
4.4.3 Electron Interferences Revisited 120
4.5 Summary 122
References 124

5 A Theoretical Description of Young-Type Interferences Following Auger Electron Emission 125
5.1 Quantum Mechanical Description 125
5.1.1 Collision Diagrams 125
5.1.2 Basic Quantum Mechanics Formalism 126
5.1.3 Application to Ion-Atom Collisions 128
5.1.4 Application to Ion-Molecule Collisions 137
5.2 A Semi-classical View of Interferences 140
5.2.1 Introduction 140
5.2.2 Basis of the Model 143
5.2.3 Preliminary Results: Angular Distribution 148
References 150

Appendix 1: System of Atomic Units 151

Appendix 2: Scattering Cross Section 155

Appendix 3: Interference Term in H_2 Ionization 157

Appendix 4: Post Collision Interaction in Atomic Collisions 163

Appendix 5: Path Interference Model 171

Appendix 6: Oscillations in the Width and Wavelength Distribution in Photon Interference Experiments 177

Appendix 7: Oscillations at Maximum of Electron Energy Distribution 183

Appendix 8: Fast Fourier Transform 191

Appendix 9: Recoil Ion Momentum and Capture Process 199

Appendix 10: Gamma and Hypergeometric Functions 203

Appendix 11: Visibility of Interference Pattern 207

Conclusion and Perspectives 211

Index 221

Chapter 1
Photon Interferences: History and Fundamental Aspects

1.1 The History of Light

1.1.1 Before Newton

1.1.1.1 Early Conceptions About Light

The conceptions of the ancient peoples about light and vision provoke a smile today, because they seem very naïve. However, one should not forget that scientific developments always stem from naïve ideas. On the one hand, the ancient peoples, by their inventiveness and imagination, laid down an essential basis for a better understanding of our world. On the other hand, strong personalities, such as Aristotle or Newton, as well as the power and influence of the Church, made questioning many of these bases difficult.

In all of the ancient civilizations, light was neither viewed as an object nor as a subject of study [1]. Light was, existed, appeared, but nothing was said about its origin or its nature [1]. Light was just one element among others, such as air, earth and fire, to name but a few. Light was necessary for life, and a stage in the Creation. Finally, light was often associated with darkness, the two existing, but separately, co-existing in order not to destroy each other.

In the ancient Greek civilization, two opposite concepts appeared. According to Democritus and Epicurus, simulacra (or ειδωλον in Greek), sort of material films composed of thin layers of atoms, streamed from the surface of objects and entered the eyes or mind, thereby causing vision and visualization [2]. The impact of these objects on our sense organs enabled us to perceive them. Plato, on the other hand, was convinced that a fire lived in the eye. This fire emanated from the eye (extramission theory), and mixed with daylight to form a transparent and homogeneous body that extended from the eye to the object [3]. With Aristotle, vision was only possible if the object acted on the eye through a material support between the eye and the object [4]. Euclid was the first to introduce mathematics to describe some of the effects of light. Observing the shadows projected by objects, he

F. Frémont, *Young-Type Interferences with Electrons*, Springer Series on Atomic, Optical, and Plasma Physics 77, DOI 10.1007/978-3-642-38479-0_1,

concluded that light propagated in straight lines, and was formed by individual rays. It was also Euclid who, after experimental observation, formulated the law of reflection.

1.1.1.2 The Arab Contribution

Optics made considerable progress from the ninth century on, essentially due to the contribution of Arab scientists, notably Al-Kindi, Ibn Sahl, and Ibn Al-Haytham, better known in the Western world as Alhazen [5]. Through experiment, they discovered some fundamental laws. They furthered Greek concepts, performed experiments on reflection and refraction, and worked on mirrors, especially those described by Archimedes. They established their own numerical tables in the case of refraction. Alhazen, known as the "Father of modern optics", proved that light, consisting of individual rays, traveled in a straight line, with finite velocities depending on the nature of the object crossed by the light.

1.1.1.3 Light in the Western World: Beginnings of a Scientific Process

During the Middle Ages, optics studies were carried out in many countries throughout Europe, notably in England, Poland, Germany, and Italy. In England, the Arab works were translated and taught. Robert Grosseteste, Roger Bacon and John Peckham were the main actors in the discovery of optics as a science [6]. Direct, reflected and refracted rays were studied. Rainbows were analyzed [7], and the seven primary colors were introduced. The lens effect was discovered. The eye was studied as an optical object. Glasses were made for the old and the shortsighted.

In 1270, the Polish friar Witelo wrote one of the earliest treatises on optics, entitled *Perspectiva*. Using a goniometer derived from a similar instrument used by Alhazen, he focused on the refraction law from the observation of rainbows, and formulated his results as a table of values of the angles of incidence and refraction. He also described the production of artificial rainbows through refraction in crystals or in bottles filled with water.

Theodoric of Freiburg, from Germany, wrote a book on rainbows, entitled *De iride et radialibus impressionibus* [8]. He explained in detail the colors and the position of the primary and secondary rainbows, as well as the reason why the secondary rainbow displays the same colors in the reverse order and has a larger radius.

Many applications of optics were carried out in Italy. For example, at the end of the thirteenth century, Alessandro della Spina was making glasses [9]. Leon Battista Alberti wrote a treatise, *Della Pittura*, on linear perspective [10]. He used his knowledge of optics to determine perspective as an instrument of artistic and architectural representation.

However, despite the efforts of these specialists, the most phenomenal breakthrough occurred at the beginning of the sixteenth century, with the use of the telescope by Galileo. There are several contenders for the title of inventor of the telescope. It is generally considered that Hans Lippershey, from the Netherlands, came up with the earliest design. Galileo first observed the Moon [11], and then went on to discover four of Jupiter's satellites. In 1676, the telescope allowed a Danish astronomer, Ole Rømer [12], to measure the speed of light for the very first time by studying one of Jupiter's moons, Io, at different times of the year. The measurement was obtained with an error of only 30 %.

During the early seventeenth century, Johannes Kepler founded the basis of a new optical science. He defined the light ray, not to be confused with light itself. In his *Astronomiae Pars Optica* [13], he investigated the formation of pictures, explained vision by refraction within the eyes, and the depth of perception by the use of both eyes. He showed that the image of an object constituted the intersection point of light rays.

At the end of the seventeenth century, René Descartes discussed the nature of light [14] using a mechanical approach. For Descartes, space was filled with matter, and light was considered as nothing more than a certain movement or action. He also attempted to derive the reflection and refraction laws through a series of analogies to the behavior of balls on surfaces. Basing his argument on philosophical considerations, he maintained that light propagated instantaneously. He also added that vision was essentially a mechanical process, with rays of light mechanically stimulating the eyes, and then these stimulations passing mechanically to the interior of the brain. Using the laws of reflection and refraction, he confirmed by calculations that the angle subtended by the edge of the rainbow and the ray passing from the sun through the rainbow's centre is 42° for the primary arc, and 52° for the secondary arc.

1.1.2 Newton and Huygens: Two Opposite Approaches

1.1.2.1 Grimaldi and Visualization of Undulations

A real revolution occurred with the discovery of the diffraction phenomenon by Francesco Maria Grimaldi in the middle of the seventeenth century [15]. Grimaldi let sunlight into a completely darkened room, through a very small slit. He inserted an opaque rod into the cone of light thus produced, and observed the shadow cast on a screen located behind the rod. He first noted that the size of the shadow was much greater than what rectilinear projection would have predicted. The shadow was bordered by alternatively bright and dark bands (fringes). This experiment contradicted the notion of an exclusively rectilinear passage of light, and created the possibility of a new mode of transmission. Diffraction constituted the first evidence of the fluid nature of light. However, Grimaldi did not discuss the notion of periodicity in the appearance of fringes.

1.1.2.2 Newton and the Corpuscular Vision of Light

In his works [16], Newton tried to describe the observable phenomena without regard to any hypothesis as to their cause. He demonstrated that, contrary to what was commonly accepted, colors were not produced by the material through which the light passed, but originated from the light itself. To do this, he carried out what he referred to as an *experimentum crucis* (crucial experiment), in which a beam of sunlight fell through a small hole onto a prism. The white light was decomposed into several colors. Separating out a blue ray, Newton demonstrated that when this ray was sent through a second prism it remained blue. However, when the entire spectrum of colors passed through the second prism, it was recomposed into white light on exiting the prism. He finally showed that the separation of light into its component colors was due to their degrees of refrangibility.

For Newton, light was composed of small massive particles, or corpuscles, whose size depended on their color. This vision was explicitly stated in one of his famous queries [17]:

> "Are not the rays of light very small bodies emitted from shining substances?"

This corpuscular theory was used to explain the reflection and refraction of light. Although reflection can be explained using corpuscles, refraction is more difficult to explain using this model because it leads to the conclusion that the velocity of light is greater in a material than in air.

Newton, on the subject of diffraction, was not opposed to the wave theory. To explain the diffraction phenomenon, he suggested that bodies acted on light at a distance to bend the rays. He attempted to link differences in refrangibility with differences in "flexibility" and the bending that may produce color fringes. He even suggested that vision might be the result of the propagation of waves in the optic nerves. In any case, wave behavior had nothing to do with the light itself, but with the medium.

1.1.2.3 Huygens and the Wave Vision

Christiaan Huygens published his works in 1690 [18]. He was not convinced by the corpuscular theory advanced by Newton. If light consisted of particles, two different light beams that cross should result in particles in all directions. But experiments proved the opposite. His wave concept was based on the experimental result by Rømer on the velocity of light. Since velocity was not infinite, light was propagated in a medium (ether). This ether consisted of uniformly elastic particles compressed very close together. Light was not an actual transference of matter but rather a "tendency to move", a serial displacement similar to a collision proceeding through a row of balls. A colliding particle would transfer its tendency to move to all particles close to the first one. Huygens therefore concluded that new wavefronts originated around each particle touched by light and extended outward from the particle in the form of hemispheres. Single wavefronts originating at single points

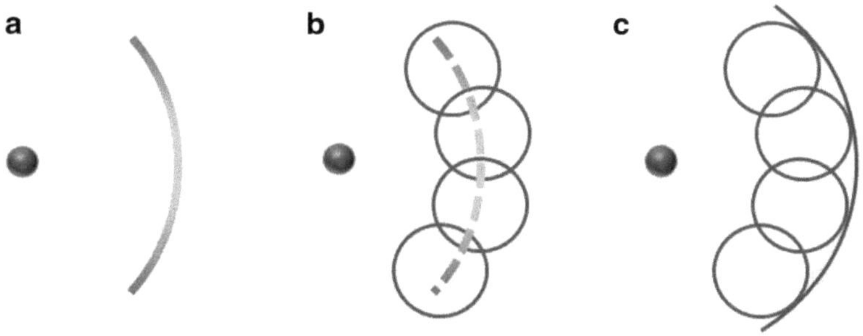

Fig. 1.1 A *spherical* wavefront produced by a light source (**a**), consisting of multiple secondary sources (**b**), and giving rise to a new wavefront (**c**)

were infinitely feeble, but where an infinite number of these fronts overlapped, there was light on the envelope encompassing all of the fronts of the individual particles (Fig. 1.1). This is "Huygens' principle".

Using this principle, Huygens was able to very easily explain reflection and refraction, as well as double refraction in Iceland spar. However, he was unable to explain colors or polarization.

1.1.3 Young: The Discovery and Formulation of Interferences

1.1.3.1 Beginnings of a Discovery

As mentioned above, Newton was not against an undulatory explanation of the effects of light, as the following example undoubtedly shows. The mixture of waves was reported for the first time from a strange observation [19]. In 1678, an employee of the East India Company, Francis Davenport, observed and reported very strange tidal patterns at a place called Batsha in the Gulf of Tonkin:

> [...] during the continuance at Batsha I have observed such an order of constancy in the course of the tides, that notwithstanding I must needs confess it different from all that ever I observed in any other Port.

In this place, there is only one tide per day, instead of two as commonly seen, with the highest tidal range coming at intervals of about 14 days, instead of the expected 15. Edmond Halley was the first to take up the matter, and said that "the effect of the moon upon the waters, in the production of the tides, [...] is the more wonderful and surprising, in that it seems different in all its circumstances from the general rules" [15]. He understood that the different effects of the tide were closely related to the Moon's position [15]. He proposed a simple formula, in which the cosine function was used to model the periodic nature of the tide.

In Book 3 of his *Principia*, Newton explained these anomalous tides [20]. *Principia* was published in 1687 and, alerted by Halley, his confident, Newton added a long paragraph providing an explanation. He suggested that the unusual ebb and flow at Batsha was the result of the combination of two separate tides flowing towards that port from different directions, one from the Indian Ocean to the South, the other from the South China Sea to the North. These tides would sometimes counteract each other, resulting in only one high and one low tide each day. Although he did not use the term *interference*, by implication Newton had hit upon the idea of wave interferences, a concept that would later play a decisive role for light and sound waves [21].

1.1.3.2 Young and the Concept of Interferences

The point we are going to develop here is one of the most important for understanding and correctly formulating analogies between photon and electron interferences. As mentioned in the introduction, many electron interference experiments are referred to as *Young-type double-slit experiments*, by analogy with the well-known experiment by Thomas Young in 1807. This analogy is erroneous, as shown a few years ago [22], and as we shall see later in more detail. It is also too restrictive, and emphasizes the ignorance of the amount of work performed by Young. Thus, it is of great importance to draw our attention to this decisive and fruitful period.

Before focusing on the nature and properties of light, Young studied medicine [23] and had "to deliver a lecture upon some subject connected with medical studies" [19]. He chose to focus on the human voice and sounds. He "collect[ed] all information relating to [the subject]" [24], and was so impressed with the resemblance between sounds and colors, that he "suspect[ed] an analogy between them".

In 1801, Young invoked interference to explain the colors of thin plates. He clearly wrote [20]:

> But the general law [...] may be very easily deduced from the interference of two coincident undulations, which either cooperate, or destroy each other, in the same manner as two musical notes produce an alternate intension and remission, in the beating of an imperfect unison.

It is remarkable, in this proposition, that the condition of interferences is only that two waves have to overlap. It does not require that the components of the waves have a common source. We place particular emphasis on this point because, as we shall see later, the electron interference experiments we mentioned previously are based on this assertion by Young, rather than on Young's 1807 experiment.

In 1804, Young carried out a preliminary experiment [25]. He made a hole in a window shutter, covered it with thick paper, and perforated it with a needle. Using a narrow slip of card, he observed the effects of a beam of sunlight on a wall. A pattern of fringes appeared, due to the combination of light waves passing on both sides of the card; more precisely, "Besides the fringes of colors on each side of the shadow, the shadow itself was divided by similar parallel fringes [...]" [21].

The famous double-slit experiment finally appeared in 1807, 6 years after Young's first assertion. "The simplest case appears to be, when a beam of homogeneous light falls on a screen in which there are two very small holes or slits" [26]. Young recognized the importance of having a common source to observe the interference fringes. The phenomenon was, according to Young, exactly the same as that found when a wave of water reaches two stones. The double-slit arrangement was used to estimate the wavelength corresponding to different colors.

1.1.4 Michelson and Morley: A Genius and Decisive Experiment

For about 100 years, light was seen as a wave. The corpuscular theory disappeared. Gradually, the wave notion took on new dimensions, and theories were developed and finalized. At the end of the nineteenth century, James Clerk Maxwell introduced five equations to describe all electromagnetic waves, including light [27]. "The propagation of undulations consists of one of these forms of energy into the other,[1] alternatively, and at any instant the amount of energy in the whole medium is equally divided, so that half is energy of motion, and half is elastic resilience." [28].

To summarize, Maxwell was convinced that a medium (ether) was necessary to transport electromagnetic waves. The question remained of the constitution of the ether. According to electromagnetic equations, the medium had to be fluid, much more rigid than steel, massless, without any viscosity, and completely transparent.

The solution to detect the presence of ether was provided by Albert Michelson and published in 1887 [28] with Edward Morley. Michelson built a very sensitive and complicated device, well-known today as the Michelson interferometer. This interferometer consisted of a light source (Fig. 1.2), delivering rays that were partly reflected and transmitted, and then returned by mirrors (1) and (2). The reflected rays were again combined and interfered. The resulting intensity was finally detected, and fringes expected to be observed.

If we suppose the ether to be at rest, the directions and distances traversed by the rays will be altered, depending on the direction of the Earth's velocity in its orbit. Therefore, displacements of the fringes can be expected. Michelson's theoretical estimation of the displacement was of the order of 0.4 fringe at maximum (dashed curve in Fig. 1.3), which is 20 times greater than that observed (full curve in Fig. 1.3).

Michelson concluded that if a relative motion between the Earth and the ether existed, it had to be small. Finally, since the concept of ether raised too many

[1] Maxwell refers to the energy responsible for the motion, and the potential energy.

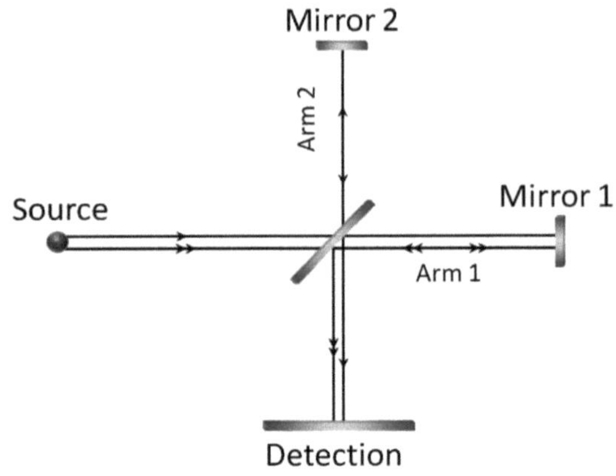

Fig. 1.2 Schematic view of the Michelson apparatus. The two rays on arms 1 and 2 originate from a unique source, and are combined and detected, after reflection on two mirrors

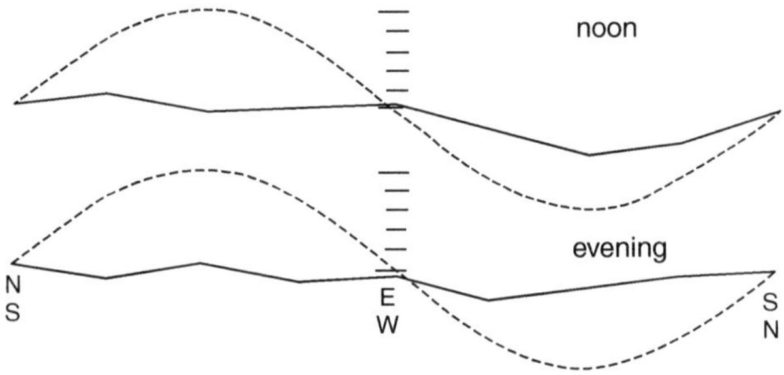

Fig. 1.3 Theoretical (*dashed curve*) and experimental (*full curve*) displacement of the fringes, for different positions of the interferometer relative to the direction of the Earth velocity, at noon (*top* of the figure) and in the evening (*bottom* of the figure)

problems, Einstein suggested that an electromagnetic wave did not need any medium to propagate, and that the speed of light was a constant, whatever the relative speed of the light source. In addition, the velocity of an object could reach the speed of light but not go beyond it.

Independently of these important conclusions, Michelson's interferometer is a reference apparatus for interferences, due to the high precision measurements. We shall see later its usefulness.

1.1.5 The Quanta Revolution and Corpuscular Vision

At the end of the nineteenth century, physics seemed to have clarified and solved all the problems.

> There is nothing new to be discovered in physics now. All that remains is more and more precise measurement.

This statement, attributed to either Lord Kelvin or Michelson, well summarizes the frame of mind of the scientists of the time. However, at least one problem still remained, which would reopen the debate. Imagine a hollow metallic sphere, pierced with a small hole, and uniformly heated. The sphere exchanges its thermal energy with both the interior of the sphere, and the exterior, in the form of electromagnetic radiations. The average wavelength will vary depending on the values of the temperature inside. More precisely, the radiation color is not unique. The outgoing radiation is formed by a superposition of many radiations with different wavelengths, weighted by a probability associated with each wavelength. If we suppose that the energy can take any value, the calculation cannot reproduce the experiment. To avoid the divergence, Max Planck used a mathematical artifact [29]. He supposed that the energy could only take integer multiples of a certain minimum value ε_o, i.e., ε_o, $2\varepsilon_o$, ... , $n\varepsilon_o$. The agreement between theory and experiment can be found on condition that $\varepsilon_o = hc/\lambda$, where c is the light velocity, λ is the radiation wavelength, and $h = 6.62 \times 10^{-34}$ Js is a constant. Thus, radiation is no longer continuous, but discontinuous, or discrete. This discovery radically changed the old conception about physics. In fact, the small constant h highlighted another problem, and became the key to new physics.

The discovery of the photoelectric effect by Heinrich Hertz in 1887 [30] brought with it the first phenomenon of the action of light on matter that wave theory was unable to interpret. Suppose light with a wavelength λ illuminates a surface. If λ is less than a certain quantity λ_{min}, electrons will be removed from the surface. If λ is greater than λ_{min}, whatever the amount of light reaching the surface, no electron will be removed. In 1905, Einstein explained this result by introducing a corpuscular theory of light. He admitted that light was composed of small bullet-like particles, called quanta, whose energies were hc/λ. Note that this corpuscular theory was only partial, since the wavelength aspect remained.

Finally, one should keep in mind that, at the beginning of the twentieth century, the question of the nature of light remained unsolved.

1.1.6 Two Crucial Experiments

1.1.6.1 Light Interferences with an Attenuated Beam

The experiment by Geoffrey Ingram Taylor in 1909 marked a turning point in the concept of the electromagnetic wave. In this experiment, Taylor proved a prediction made by Joseph John Thomson two years earlier, a theory in which light energy was distributed unevenly over the wavefront [31].

> The theory considered in this paper – that the electric field is made up of a number of discrete units – is one which naturally suggests itself, if we use the conception of tubes of electric force for representing the state of the electric field. [32]

There were regions of maximum energy widely separated by large undisturbed areas, in contrast with the uniform wavefront of the usual electromagnetic theory. If Thomson's theory was correct, the intensity of light in a diffraction pattern would be modified when the source intensity was considerably reduced.

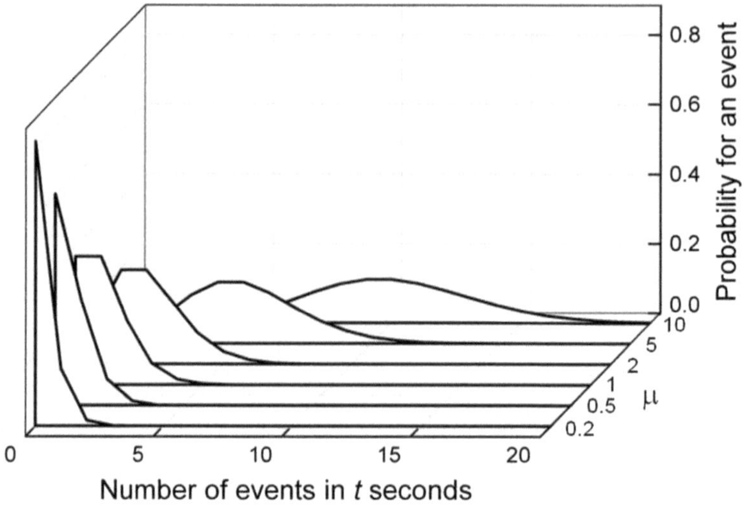

Fig. 1.4 Probability $p(k)$ as a function of k for five values of μ

Taylor used as a source a narrow slit placed in front of a gas flame [33]. Various screens were installed to reduce the intensity. The largest time of exposure was about 2,000 h, or 3 months. As observed by Taylor, "in no case, was there any diminution in the sharpness of the pattern".

1.1.6.2 Interferences with Single Photons

Taylor's experiment was considered for many years as a single photon experiment. Paul Dirac himself concluded from this experiment that *a photon can interfere with itself* [34]. In fact, as it was shown 80 years later, Taylor's experiment was not, strictly speaking, a single photon experiment. A light beam, attenuated or not, contains a number N of elementary particles per second. Suppose that there are on average μ events during a time t, the probability to detect k events in the same time is defined by the Poisson probability law:

$$p(k) = e^{-\mu} \frac{\mu^k}{k!} \tag{1.1}$$

The result for $p(k)$ is presented in Fig. 1.4 as a function of k, for five values of μ. If μ is much greater than one, the probability has the form of a Gaussian curve. If μ diminishes, the probability to obtain two or more than two events also decreases, but is never zero. In other words, Taylor's experiment only shows that, on average, one photon is detected. But the probability for the detection of two or more photons cannot be neglected.

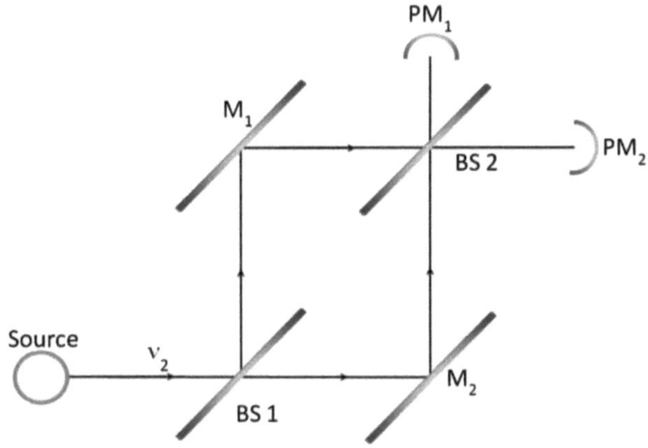

Fig. 1.5 Mach-Zender type interferometer. The light is split (BS1), reflected on Mirrors 1 and 2, and finally detected on PM1 and PM2

To show interferences with single photons, one must therefore be sure that only one photon at a time crosses the interferometer. The only solution consists of a randomly oriented photon source [35]. Photons are obtained by excitation with two lasers. After excitation, one atom of the crystal de-excites, emitting a group of two photons by radiative cascade. The first photon, with frequency ν_1, is used as a start for the detection of the second photon, of frequency ν_2, forming the one photon light pulse. Suppose now that two groups of photons are created at the same time. Because the emission is spontaneous, the photons are emitted in any direction, so that the detection of the second group is unlikely to occur.

The interference experiment was carried out in 1985 using a Mach-Zehnder interferometer. The light was first split (BS 1), then reflected on two mirrors (M_1 and M_2), and recombined using a second beam splitter. Finally, the resulting light was detected on two photomultipliers (PM_1 and PM_2), and the intensity recorded as a function of the path difference, controlled by the displacement of the mirrors (Fig. 1.5). The result is shown in Fig. 1.6, for four different times. The interference fringes are clearly visible, showing that one photon can really interfere with itself.

1.2 Characteristics of Interferences

1.2.1 Conditions for Interferences

To fix the problem, let us consider the case of two sources S_1 and S_2, delivering two electromagnetic waves, whose amplitude are, at a given point M on a screen:

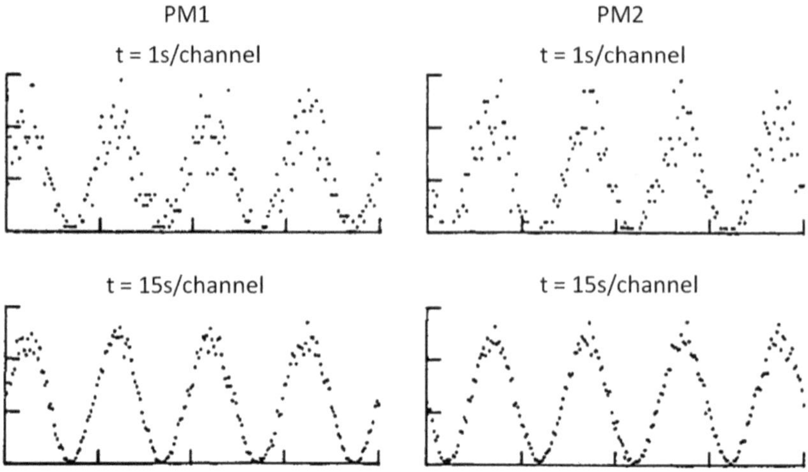

Fig. 1.6 Number of counts detected on PM1 (*left side*) and PM2 (*right side*), as a function of the path difference [36]

$$
\begin{aligned}
y_1(t) &= a_1 \cos\left(\frac{2\pi c}{\lambda_1}\left(t - \frac{S_1 M}{c}\right) + \varphi_1\right) \\
y_1(t) &= a_2 \cos\left(\frac{2\pi c}{\lambda_2}\left(t - \frac{S_2 M}{c}\right) + \varphi_2\right)
\end{aligned} \tag{1.2}
$$

where λ_1 and λ_2 are the wavelengths, and φ_1 and φ_2 are phases. Let us suppose, to simplify matters, two independent sources of the same intensity, i.e., $a_1 = a_2 = a_o$. If these sources originate from the spontaneous emission of light, the phase difference $\Delta\varphi = \varphi_1 - \varphi_2$ will depend on time, since φ_1 and φ_2 also depend on time. The total intensity is given by:

$$
I = I_1 + I_2 + \sqrt{I_1 I_2}\cos\left(\left(\frac{2\pi c}{\lambda_1} - \frac{2\pi c}{\lambda_2}\right)t + \frac{2\pi S_1 M}{\lambda_1} - \frac{2\pi S_2 M}{\lambda_2} + \Delta\varphi\right) \tag{1.3}
$$

The total intensity I is averaged on time, giving rise to $I = 2\,I_o$, where I_o is the intensity of the primary source. No interference pattern is observed. To observe interference fringes, $\Delta\varphi$ has to be a constant over time. In this case, $\Delta\varphi$ does not play any role and, with the condition that the two wavelengths are both equal to the same value λ, the total intensity is:

$$
I = 4 I_o \cos^2 \frac{\Delta\varphi}{2} \tag{1.4}
$$

where $\Delta\varphi = 2\pi L/\lambda$ and $L = S_1 M - S_2 M$

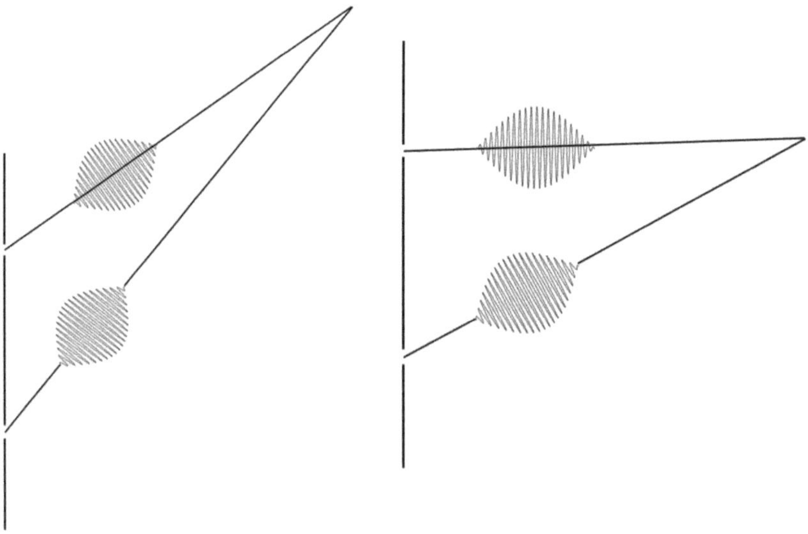

Fig. 1.7 Light is characterized by its dispersion in wavelength. On the *left*, the two wave packets do not overlap. Interferences do not occur. On the *right*, however, the two wave packets overlap, so that interferences play a role

1.2.2 Coherence in Time and in Space

To obtain interferences, waves have to overlap, even partially (Fig. 1.7). If the wave packet has a dispersion time equal to τ, interference will occur if $\tau > L/c$, where L, derived from the interference device, is the optical path difference defined in the previous paragraph.

The coherence time τ_c is the minimum time required for interferences to begin to play a role, and is defined as $\tau_c = 1/\Delta\nu = \lambda^2/\Delta\lambda$. The coherence length is thus $L_c = c\tau_c = c\lambda^2/\Delta\lambda$, and has to be less than L.

These concepts, valid for light, are also utilized for massive particles, and will be discussed in the case of single-electron interferences.

1.2.3 Contrast and Visibility of Fringes

1.2.3.1 Definition

Contrast is related to the degree of coherence. In practice, contrast can be defined as

$$C = \frac{I_{\max} - I_{\min}}{I_{\max} + I_{\min}} \tag{1.5}$$

Fig. 1.8 Contrast C as a function of the minimum of intensity I_{min}

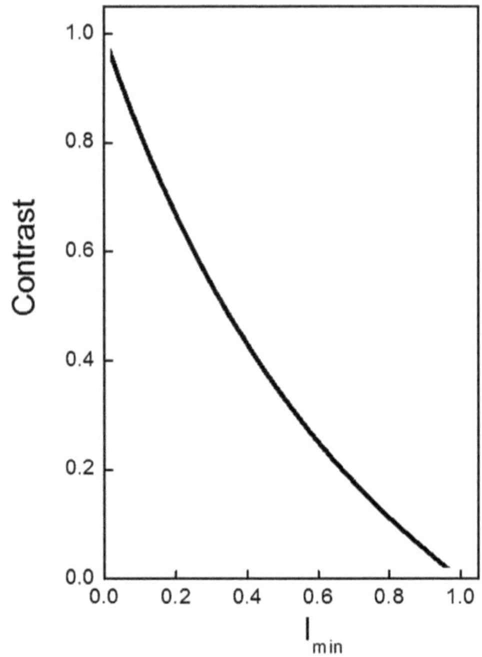

where I_{max} and I_{min} are, respectively, the maximum and the minimum values for the intensity. Using this formula, it is easy to see that C ranges from 0 to 1 (Fig. 1.8.). The contrast is 0 when $I_{max} = I_{min}$ (no visible fringes), and is equal to 1 when $I_{min} = 0$ (maximum visibility).

1.2.3.2 Limitations of the Contrast

As we shall see later in Chap. 4, obtaining a sufficiently pronounced contrast in electron interferences following ion molecule collisions is not straightforward. At least three parameters systematically limit the contrast:

- Electron spectra are distributed over a large range of energies. The mean energy is thus defined with a width ΔE. Since the energy corresponds to a wavelength λ, the source emitted is polychromatic. This limitation is characteristic of a partial time coherence (Sect. 1.2.3.2.1).
- The emitted electron is not localized in space. When the electron scatters on the slits formed by two protons, the interference pattern will depend on its initial position. In the case of light interferences, this means that the source is not punctual, but extended (Sect. 1.2.3.2.2).
- In addition to electron emission that gives rise to Young-type interferences, the emission may originate from other processes, such as direct ionization. These processes generate a large background. Similarly, in addition to the light that

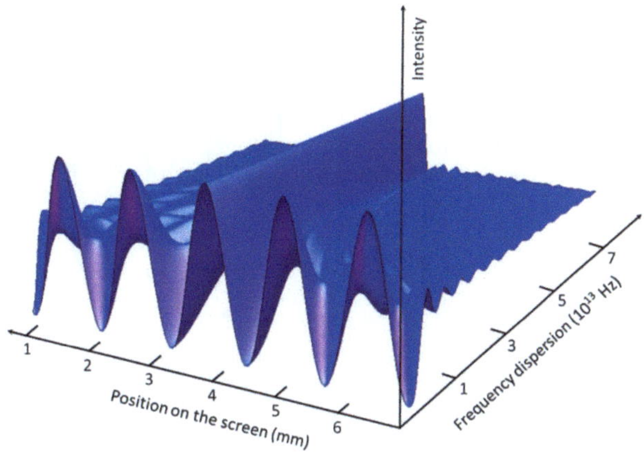

Fig. 1.9 Interference pattern produced by a polychromatic source. When the width of the frequency (or the wavelength) distribution increases, the contrast decreases

crosses the holes, part of it passes on each side of the interferometer and directly illuminates the screen (Sect. 1.2.3.2.3).

Polychromatic Source

With a monochromatic source ($\Delta\lambda = 0$), the interference pattern consists of a series of well-defined maxima and minima in intensity (Fig. 1.9). When $\Delta\lambda$ increases, the interference pattern is modified. The contrast is unchanged in the center of the figure, and strongly decreases as the position from the center increases.

The diminution of the contrast can be explained by the following calculation. To visualize the situation, let us take the double-slit device. The holes are separated by the distance a and the observation screen is located at a distance D from the holes. Suppose a rectangular spectral profile of the light. The source emits light whose frequency ν is in the $\nu - \Delta\nu/2$ to $\nu + \Delta\nu/2$ range, with the same intensity in this range. The elementary intensity received on the screen, due to the superposition of the waves, is

$$dI = 2I_o\left(1 + \cos\frac{2\,\pi xa}{cD}\nu\right)d\nu$$

The total intensity is obtained by the integration of the previous expression over the frequency.

$$I = 2I_o\left(1 + \frac{\sin\frac{\pi xa}{cD}\Delta\nu}{\frac{\pi xa}{cD}\Delta\nu}\underbrace{\cos\frac{2\pi xa}{cD}\nu}_{interference\ \ term}\right) \tag{1.6}$$

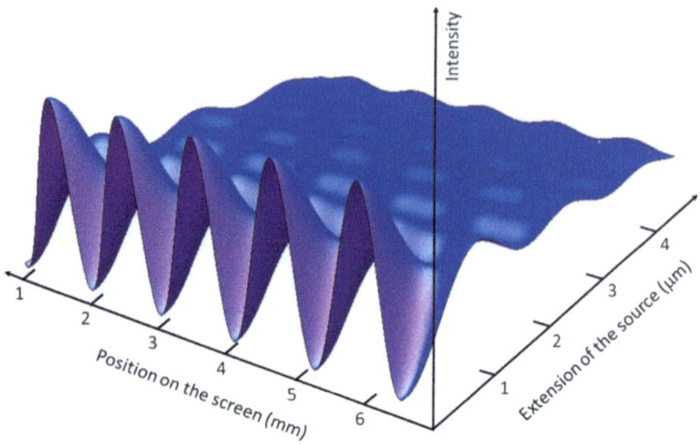

Fig. 1.10 Interference pattern produced by a spatially extended monochromatic source. When the extension of the source b increases from 0 to 4, the contrast decreases

The quantity $\sin\frac{\pi xa}{cD}\Delta\nu/\frac{\pi xa}{cD}\Delta\nu$ is an additional term compared to the classical interference term. Its consequence is to decrease the maximum of intensity when one moves away from the center, thus decreasing the contrast.

Spatially Extended Source

When the source is spatially extended, the argument is similar to the previous one. Suppose that the two Young holes are illuminated by a light source of width b. This source can be considered as the incoherent combination of infinite punctual sources, each of them giving rise to an individual intensity

$$dI = 2I_o\left(1 + \cos\frac{2\pi}{c}\left(\frac{xa}{D} + \frac{Xa}{d}\right)\nu\right)dX$$

where d is the distance between the primary source and the holes, and X is the distance between the center of the source and any point of it.

The total intensity received on the screen is the sum of the entire individual intensities dI.

$$I = 2I_o\left(1 + \frac{\sin\frac{\pi ab}{\lambda D}}{\frac{\pi ab}{\lambda D}}\underbrace{\cos\frac{2\pi xa}{\lambda D}}_{interference\ term}\right) \tag{1.7}$$

The total intensity is weighted by $\sin\frac{\pi ab}{\lambda D}/\frac{\pi ab}{\lambda D}$, so that the total intensity is decreased by a constant factor (Fig. 1.10). The contrast is

$$C = \frac{\sin\frac{\pi ab}{\lambda D}}{\frac{\pi ab}{\lambda D}\left(2 + \sin\frac{\pi ab}{\lambda D}/\frac{\pi ab}{\lambda D}\right)} \tag{1.8}$$

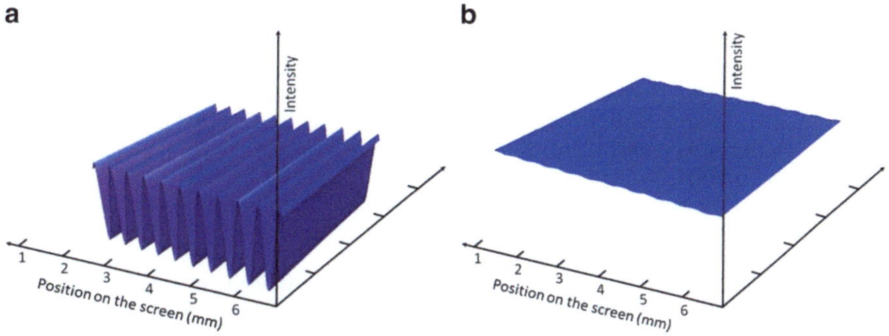

Fig. 1.11 Interference pattern produced by two punctual sources without any background (*left-hand* side of the figure) and superimposed onto a diffuse background (*right-hand* side of the figure)

The greater the spatial dispersion b of the source, the lower the contrast.

Combination of Two Secondary Sources and a Diffuse Background

In the case of two punctual sources associated with a diffuse background with intensity I_d, part of the light scatters on the holes, while another part passes on each side of the sheet, without interfering. The latter gives rise to a uniform intensity on the screen, combined with the interference pattern (Fig. 1.11). The greater the background, the lower the contrast, given by

$$C = \frac{I_{\max} - I_{\min}}{I_{\max} + I_{\min} + 2I_d} \tag{1.9}$$

References

1. Radin, P.: The Winnebago Indians, in thirty-seventh annual report. Bureau of American Ethnology, Washington, D.C. (1923)
2. Diels, H., Kranz, W.: Die Fragmente der Vorsokratiker, 6th edn. Weidmann, Berlin (1951)
3. Plato: Timaeus, (translated by B. Jowett), Easy Reading Series. Forgotten books
4. Aristotle: De Anima, II 7, Belles Lettres (1996)
5. Lindberg, D.C.: Alhazen's theory of vision and its reception in the west. Isis **58**, 321–341 (1967)
6. Southern, R.W.: Robert Grosseteste: the growth of an English mind in medieval Europe. Oxford (1986)
7. Lindberg, D.C.: Theories of vision from Al-Kindy to Kepler. Isis. **81**, 464–491 (1990)
8. Grant, E.: A source book in medieval science. Harvard University Press, Cambridge (1974)
9. Rosenthal, J.W.: Spectacles and other vision Aids: a history and guide to collecting. Norman Publishing, San Francisco (1994)

10. Alberti, L.B.: De la statue et de la peinture. Translated from the Latin into French by C. Popelin. Lévy éditeur, Paris (1869)
11. Galileo, G.: Sidereus nuncius. Translated and prefaced by Albert Van Helden. University of Chicago Press, Chicago/London (1989)
12. Ole Romer: (30 September 1677), *Lettre no 2104*. In: Bosscha, J. (ed.) Œuvres complètes de Christiaan Huygens (1888–1950). Tome VIII: Correspondance 1676–1684. Martinus Nijhoff, The Hague (1899)
13. Kepler, J.: Astronomiae pars Optica, a study on optics and the human eye. Partial translation into French: Paralipomènes à *Vitellion*, (translated by Catherine Chevalley) J. Vrin, Paris (1980)
14. Descartes, R.: Oeuvres complètes: Tome 3, Discours de la Méthode suivi de La Dioptrique, Les Météores et la Géométrie, Gallimard, Paris (2009)
15. Grimaldi Francesco Maria: Physico mathesis de lumine, coloribus, et iride, aliisque annexis libri duo. Vittorio Bonati, Bologna (1665)
16. Newton, I.: Opticks, or a treatise of reflections, refractions, inflexions and colours of light. Printed for W. Innys at the West End of St Paul's, London (1730)
17. Newton, I.: Optiks, 29th query
18. Huygens, C.: Traité de la lumière. Pieter van der Aa, Leiden (1690)
19. Halley, E.: An account of the course of the tides at Tonqueen in a letter from Mr. Francis Davenport, July 15, 1678. With the theory of them, at the Barr of Tonqueen. Phil. Trans. R. Soc. Lond **13**, 677 (1684)
20. Newton, I.: The Principia, Book III. University of California Press, Berkley (1999)
21. Cartwright, D.E.: The Tonkin tides revisited. Notes Rec. R. Soc. Lond. **57**, 135 (2003)
22. Barrachina, R.O., Žitnik, M.: Young's interference effect in the autoionization of atoms colliding with molecules. J. Phys. B **37**, 3847 (2004)
23. Mollon, J.D.: The origins of the concept of interference. Phil. Trans. R. Soc. Lond. A **360**, 807 (2002)
24. Young, T.: A syllabus of a course of lectures on natural and experimental philosophy. Royal Institution, London (1802)
25. Young, T.: The Bakerian lecture, experiment and calculation relative to physical optics. Phil. Trans R. Soc. Lond., Royal Society of London, 1–16 (1804)
26. Young, T.: A course of lectures on natural philosophy and the mechanical arts. J. Johnson, London (1807)
27. Maxwell, J.C.: A dynamical theory of the electromagnetic field. Phil. Trans. R. Soc. Lond. **155**, 459 (1865)
28. Michelson, A.A., Morley, E.W.: On the relative motion of the Earth and the Luminiferous ether. Am. J. Sci. **203**, 334 (1887)
29. Planck, M.: On the law of distribution of energy in the normal spectra. Ann. Phys. **4**, 553 (1901)
30. Hetz, H.R.: Ueber einen Einfluss des ultravioletten Lichtes auf die electrische Entladung. Ann. Phys. **267**, 983 (1887)
31. Thomson, J.J.: Proc. Camb. Phil. Soc. **14** (1907)
32. Thomson, J.J.: On a theory of the structure of the electric field and its application to Röntgen radiation and light. Phil. Mag. **19**, 238–247 (1910)
33. Taylor, G.I.: Interference fringes with feeble light. Proc. Camb. Phil. Soc. **15**, 114 (1909)
34. Dirac, P.A.M.: Quantum mechanics. Oxford University Press, London (1958)
35. Grangier, P., Roger, G., Aspect, A.: Eur. Phys. Lett. **1**, 173 (1986)
36. Aspect, A., Roger, G., Grangier, P.: J. Optics **20**, 119 (1989)

Chapter 2
Interferences with Massive Particles

2.1 The De Broglie Wave-Particle Duality

2.1.1 The Concept of Wave-Particle Duality

Parallel to the introduction by Einstein of the concept of quanta for light, Niels Bohr showed that the quanta idea could be extended to atoms [1]. Coming back to the model of an atom introduced by Ernest Rutherford, where the atom consisted of a "positively charged nucleus surrounded by a system of electrons kept together by attractive forces from the nucleus", Bohr pointed out "difficulties of a serious nature arising from the apparent instability of the system of electrons". It is well established that, when a charged particle accelerates, radiation is emitted. Thus, the electron will lose energy and follow orbits of smaller and smaller sizes. To avoid these difficulties, Bohr referred to Planck's theory and forced the orbits to be stable, with the electrons emitting no radiation. An atom was now considered as a microscopic solar system. At such a small scale, "the electrons and other massive corpuscles do not follow [. . .] classical mechanics laws, but can have defined motion states [. . .] that satisfy some quanta conditions" [2].

De Broglie pointed out the appearance of integers in wave theory. He therefore saw "an indication in favor of the idea that, for electrons and other corpuscles, a wave-particle duality exists, as for photons and light waves" [2]. For him, it was obvious that a general synthesis needed to be done, applicable to light and matter. In 1924, de Broglie put forward the brilliant and revolutionary hypothesis that one could associate with a particle of momentum p a wave of frequency ν or wavelength λ, related to p by $p = h/\lambda$, where h is the Planck constant [3].

During his life, de Broglie's problem was to combine the two concepts, one statistical (wave), the other individual (particle). He tried, for example, to define a spatial-temporal model of particles of matter or light which involved hidden variables to explain their corpuscles and wave properties. He introduced two kinds of wave functions to solve the problem (double solution theory) [4]. In this work, he postulated that any solution φ of non-linear wave equations was

F. Frémont, *Young-Type Interferences with Electrons*, Springer Series on Atomic, Optical, and Plasma Physics 77, DOI 10.1007/978-3-642-38479-0_2,

accompanied by a solution u with a singularity (corpuscle) that had the same phase as φ. In fact, as shown by some of his colleagues, the combination of both solutions induced mathematical complications. Thus, one possible solution would be to say that, depending on circumstances, a particle behaves like either a corpuscle or a wave.

2.1.2 The First Evidence for Wave Behavior of Massive Particles

To prove wave-particle duality for electrons, "it was natural that the first application should be to the atom. No thought was given, after de Broglie's assumption, to electrons in free flight" [5]. Work began in 1919 by analyzing the energy of secondary electrons after they scattered on metal. It was discovered purely by accident that the intensity of elastic scattering varied with the orientation of the scattering crystal. The first to draw attention to this was Walter M. Elsasser, who pointed out in 1925 [6] that a demonstration of diffraction would establish the physical existence of electron waves. Elsasser wondered whether Davisson and Kunsman maxima and minima were diffraction phenomena similar to those produced by X-rays penetrating crystals [7]. He argued that anomalous peaks in the intensity of electrons could be evidence of electron diffraction. His publication was reviewed by Einstein who remained skeptical, but Elsasser's conclusions were found to be important for physicists such as Werner Heisenberg and Max von Laue.

The search for diffraction beams began at the end of 1926. A beam of electrons was directed against a crystal of nickel (Fig. 2.1).

A collector designed to accept only elastically scattered electrons could be moved in an arc of a circle around the crystal. The crystal itself could be rotated about the axis of the incident beam.

As shown in Fig. 2.2, the angular distribution for the intensity at a fixed azimuth of the crystal presents a maximum at a particular primary-beam voltage V (54 V in Fig. 2.2).

In Fig. 2.3, the wavelength derived from the angle of diffraction is represented as a function of $1/\sqrt{V}$ (points), and compared with the relation $\lambda = h/p$ (straight line). This remarkable agreement confirmed the wave behavior of electrons.

2.2 Interferences with Massive Particles

2.2.1 Interferences with Neutrons

In 1947, 15 years after the discovery of the neutron by James Chadwick, Enrico Fermi and Leona Marshall Libby systematically studied interferences produced by

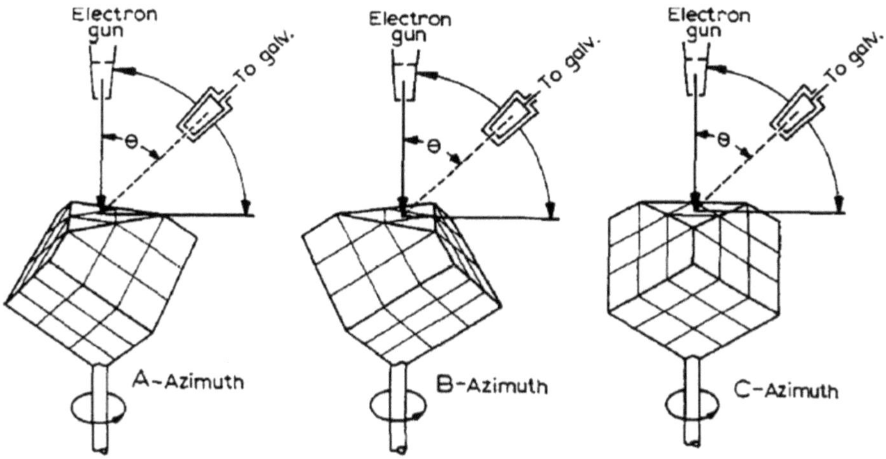

Fig. 2.1 Schematic view of the Davisson and Germer set-up, showing the arrangement of the primary electron beam, the crystal and the collector [5]. The crystal can be rotated around the azimuthal axis, and the detector can also be rotated in a plane

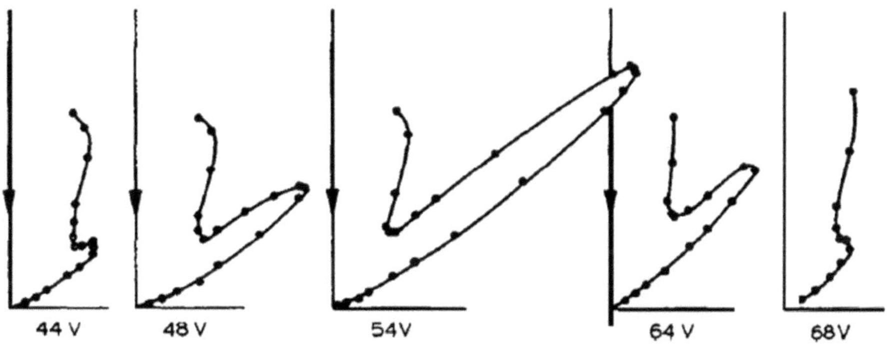

Fig. 2.2 Polar diagram showing intensity of elastic scattering in the A-azimuth as a function of the angle θ, for a series of primary-voltage beam [5]

slow neutrons scattering on various elements, such as Li, Be or C [8]. Their goal was to determine the phase change of the scattered neutron wave for a large number of elements. Their experimental set-up is schematically represented in Fig. 2.4. A beam of non-monochromatic thermal neutrons emerges from a hole containing a long collimator in the thermal column of the Argonne heavy-water pile. This beam falls on the $1, 0, 0$ plane of a large CaF_2 crystal. Neutrons of energies satisfying the Bragg condition are reflected to form a beam that is used for investigating various crystal specimens. The beam contains a main component of neutrons of 0.15 nm wavelength, plus a small fraction of neutrons of energy four times as large due to the second-order reflection on the CaF_2 crystal. The second-order component is fairly insignificant. Therefore, one can consider the beam reflected on the calcite to

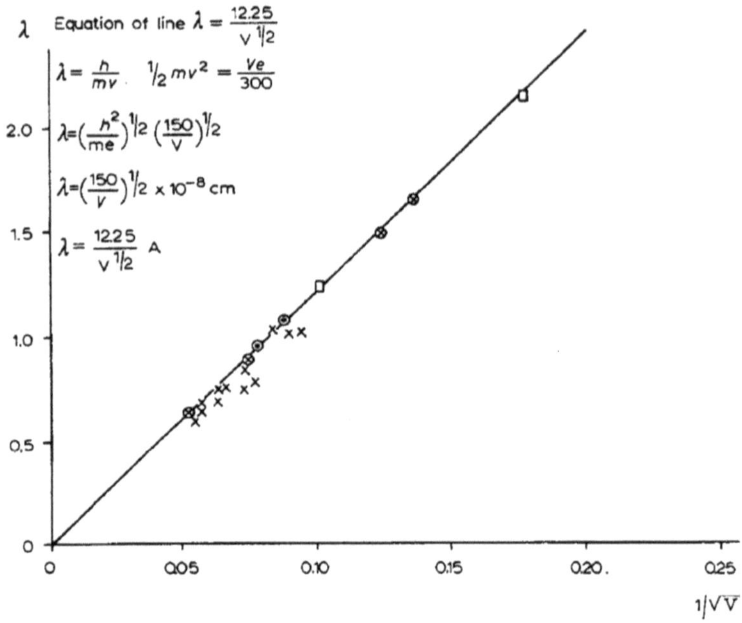

Fig. 2.3 Comparison between theoretical relation $\lambda = h/p$ and various experimental results (points) [5]

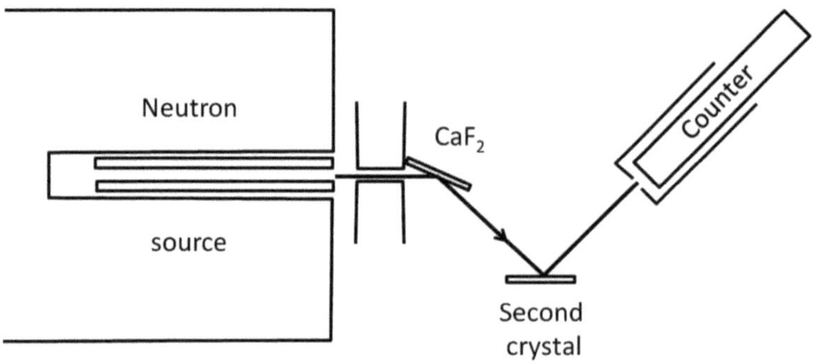

Fig. 2.4 Schematic view of the neutron interferometer set-up [8]

be approximately monochromatic. This monochromatic beam of neutrons falls on a plane of the second crystal, which is mounted on a rotating table. The neutrons undergo a second Bragg reflection, and are finally detected by a counter.

The scattering cross section σ is, in the first approximation, linked to the scattering length a by the relation

$$\sigma = \pi |a|^2 \tag{2.1}$$

A simple formula can be obtained in the case that scattering is due to the effect of a single resonant level. It is shown that a is proportional to the de Broglie wavelength λ associated with the energy of the primary neutrons. The total reflection of neutrons on mirrors is theoretically expected to occur at very small glancing angles. The limiting glancing angle is given, with very good approximation, by

$$\theta_o = \lambda \sqrt{Na/\pi} \tag{2.2}$$

where N is the atom density in the mirror. The agreement between the last formula and experimental angles is found to be good, with a relative uncertainty of less than 10 %, showing the wave behavior of the neutrons.

To observe maxima and minima in intensity, as observed for light, other set-ups were used. In 1974, a neutron interferometer was built (Fig. 2.5) with Si crystals as diffraction gratings [9]. The geometry of the device required the use of thermal neutrons to directly observe interference fringes. The first crystal played the role of beam separator. The second crystal recombined the transmitted and scattered beam, inducing interferences.

An aluminum plate, the angle of which was variable with respect to the neutron beam, was introduced to induce a path difference between the two beams. This path difference gave rise to interferences that could be detected. Figure 2.6 shows an example of the well-defined oscillations that are the signature of neutron interferences.

2.2.2 *Interferences with Atoms*

2.2.2.1 Double-Slit Experiment

Interferences with atoms were observed in the early 1990s [10]. The major problem was fringe spacing, i.e., the distance between bright and dark fringes. Using atoms at ambient temperature, the wavelength was less than one nanometer, and the fringes were therefore difficult to observe. On the contrary, if the atoms were cooled, their velocity decreased, and the wavelength increased. For example, for a velocity of the order of 100 m/s, the wavelength increased by a factor of ~100, inducing fringe spacing that could reach a few tenths of millimeters.

A supersonic jet of He crossed an interferometer, constituted by two slits separated by a distance of 8 μm (Fig. 2.7). Scattered atoms were detected by displacing a detector every 2 μm. The result of the detection is shown in Fig. 2.8 for a wavelength of 0.103 nm. Maxima and minima in intensity are visible, with a fringe spacing of 8.4 μm, in excellent agreement with theoretical predictions.

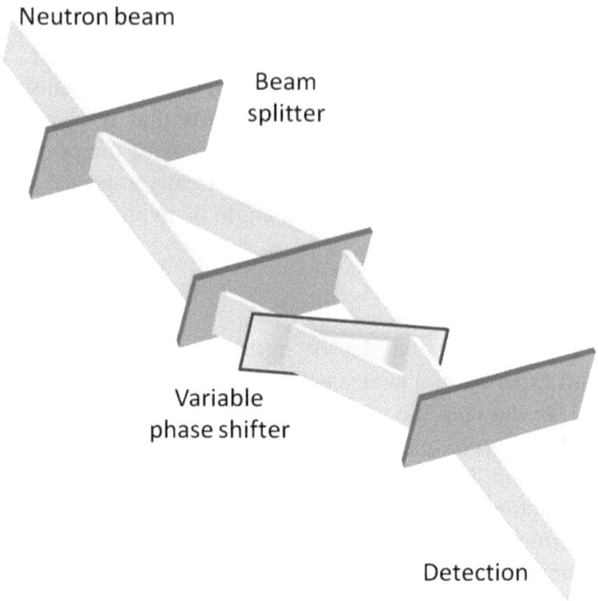

Fig. 2.5 Experimental set-up used to observe interferences produced by thermal neutrons

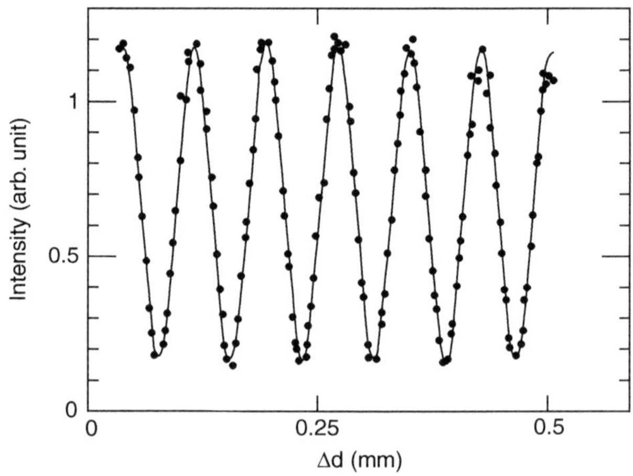

Fig. 2.6 Number of detected neutrons as a function of the path difference between scattered and transmitted beams. The oscillations are the signature of interferences

2.2.2.2 Three-Grating Experiment

To allow the largest interference signal for a given beam brightness, three-grating white-fringe geometry was used. In the experiment described in 1991 [11] for example, the atom beam consisted of a supersonic jet of sodium atoms (Fig. 2.9), giving rise to monochromaticity $\Delta v/v = 12\ \%$, with $v = 1{,}000$ m/s, inducing a wavelength of 16 pm. The diffraction gratings (G_1, G_2, and G3 in Fig. 2.9) consisted of arrays of slots in a silicon nitride membrane. The interferometer had three 400 nm-period gratings mounted apart on separate stages. The zeroth- and

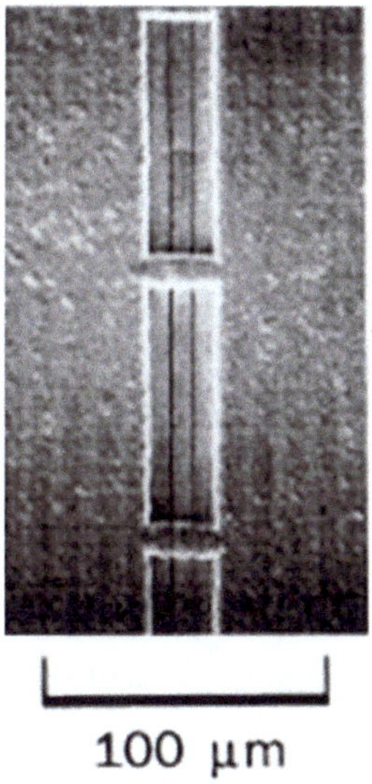

Fig. 2.7 Scanning-electron microscope picture of the transmission structure

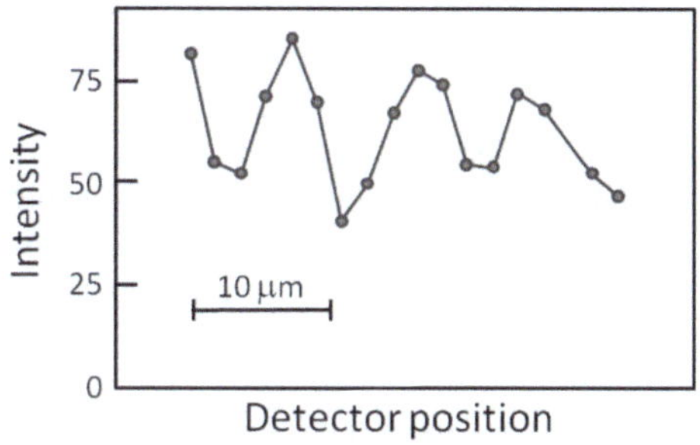

Fig. 2.8 Intensity profile in the detector plane as a function of the detector position

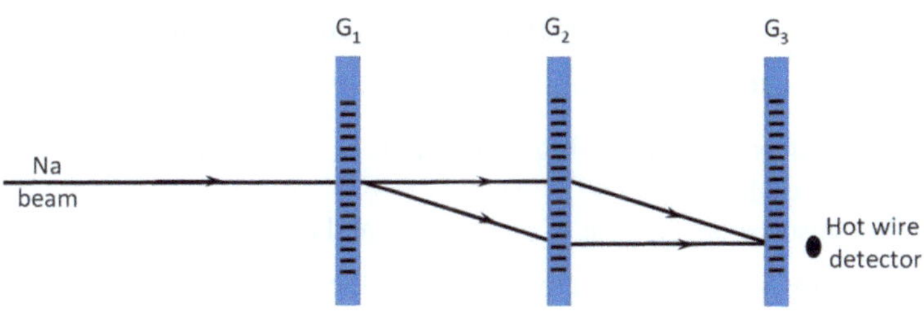

Fig. 2.9 Scheme of the three-grating interferometer [11]. The interferences are produced by a Na beam crossing the gratings G_1 and G_2

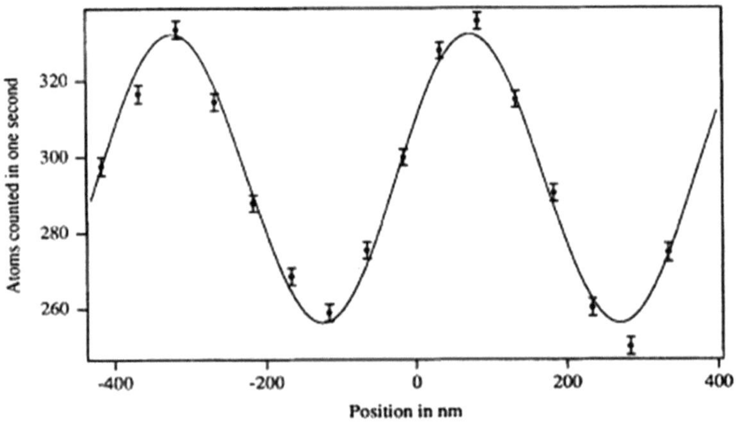

Fig. 2.10 Intensity as a function of the detector position. Experiment: *solid points*. Theory: *full curve*

first-order beams from the first grating struck the second one where they were diffracted in the positive and negative first orders to converge at the third grating. The first two gratings formed an interference pattern in the plane of the third grating, which acted as a mask to sample this pattern. A hot wire detector was moved in order to determine the number of counts as a function of the detector position.

The intensity, i.e., the number of scattered atoms, was thus recorded as a function of the detector. The experimental result is represented in Fig. 2.10 (solid points) as a function of the position. Well-defined oscillations can be observed, in very good agreement with the theoretical calculations (full curve).

2.2.2.3 Ramsey-Bordé Interferometry

The Ramsey-Bordé interferometer is analogous to the Mach-Zehnder interferometer for light (cf. Chap. 1, Sect. 1.1.6.2) which uses separators and mirrors. The principle of the beam splitter (separator) is the following. Suppose that one atom has two ground state hyperfine sub-levels $\left|\vec{k}, g_1, 0\right\rangle$ and $\left|\vec{K}, g_2, 0\right\rangle$, where 0 is the number of transferred photons (Fig. 2.11) and $\vec{k}$ and $\vec{K}$ are the atom momenta associated with each state [12].

The atom absorbs one photon of energy $\hbar\omega_l$ and emits by stimulated emission a photon of energy $\hbar\omega_2$. Thus the atom can be in the ground state $\left|\vec{K}, g_2, 0\right\rangle$ with a given probability. The absorption and emission of the two photons is accompanied by a change in the momentum $\vec{k}$, which is the sum of the two momenta $\vec{k_1}$ and $\vec{k_2}$ due to each photon (Fig. 2.12).

The Ramsey-Bordé interferometer is composed of several beam splitters similar to the one described above. An example is given in Fig. 2.13 [13]. The probability of absorption and stimulated emission is chosen to be 50 % in each interaction zone

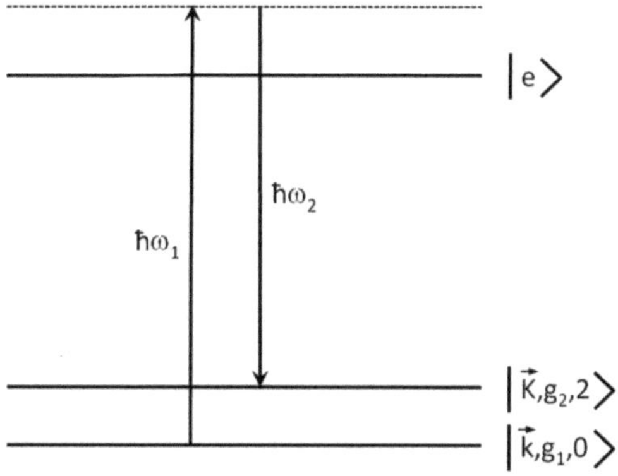

Fig. 2.11 Principle of atom diffraction using a Raman process [12]. The atom absorbs a photon of energy $\hbar\omega_1$, and emits a photon of energy $\hbar\omega_2$ by stimulated emission

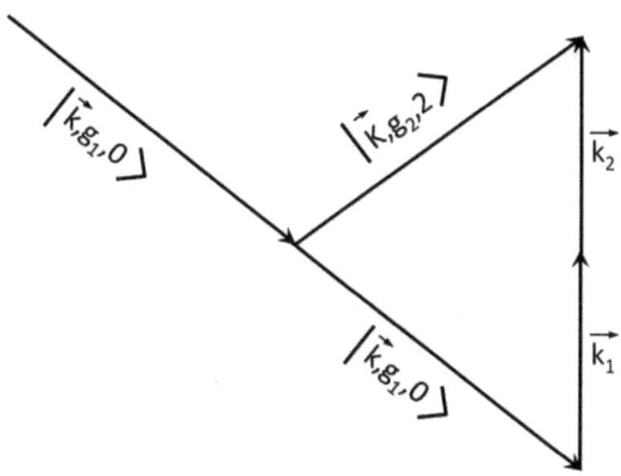

Fig. 2.12 Momentum conservation during photon absorption and emission. The initial and final momenta are $\overrightarrow{k}$ and $\overrightarrow{K}$, respectively [12]

by applying the proper laser intensity and interaction time. The first interaction (first arrow in Fig. 2.13), for example, produces the superposition $1/\sqrt{2}\left(\left|\overrightarrow{k}, g_1, 0\right\rangle + \left|\overrightarrow{k} + \overrightarrow{k_1}, e, +1\right\rangle\right)$, with a phase depending on the laser.

Two classes of interferometers can be defined [14].

- In open interferometers (e.g., Young's double-slit interferometer), a single beam divider produces a coherent superposition of waves. The interference pattern which results from the superposition of several outgoing wavefronts is explored directly by moving a detector in space or in time.
- In a closed interferometer, the outgoing waves are re-superposed by a second beam splitter. The interference pattern is observed by scanning the phase difference between the two beam splitters.

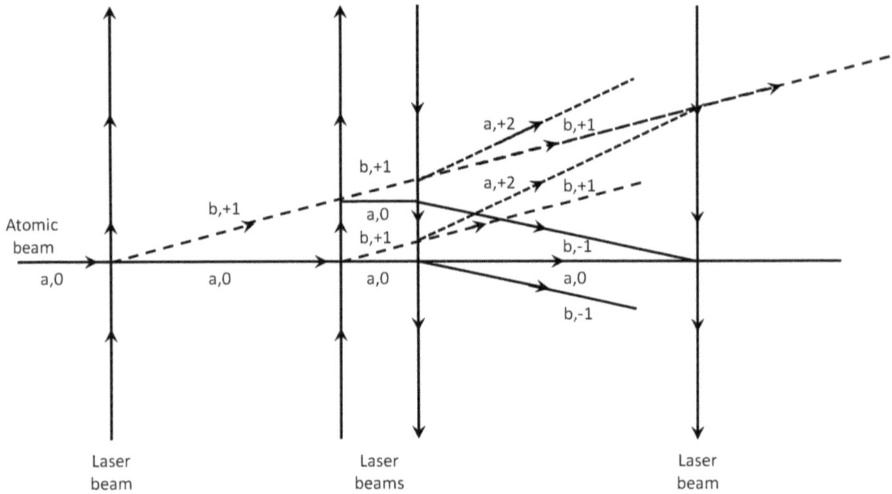

Fig. 2.13 Ramsey-Bordé type interferometer. The atom is initially in state *a,0*. The use of lasers can produce intermediate or final states labeled *b,+1*, *b,−1* and *a,+2*

The Ramsey-Bordé interferometer is particularly used for high-precision measurements of gravity, polarizability measurements, the determination of the effects of a magnetic field, and the analysis of inertial effects.

2.2.3 Interferences with Small Molecules

Since interferometry is applied to atoms, it is also well suited for small molecules. In 2000, for example, interference pattern was studied for diatomic molecules He_2. In this molecule's fundamental state, its atoms are separated by ~5 nm, which is much greater than for most molecules. These molecules are produced from an atomic source (Fig. 2.14) and then separated by beam diffraction.

The grating consists of slits of a width of 500 nm, separated by 500 nm. The trajectories of the diffracted elements depend on their mass, so that these elements can be analyzed by mass spectroscopy. Figure 2.15 shows a typical spectrum obtained after diffraction, for temperatures T of the source varying between 4.5 and 65 K. One can see first that the dimer production is weak (~1,000 times less than that of He atoms). The angular distribution reveals series of maxima and minima in intensity. When T decreases, the energy decreases, so that the wavelength and thus the fringe spacing increase.

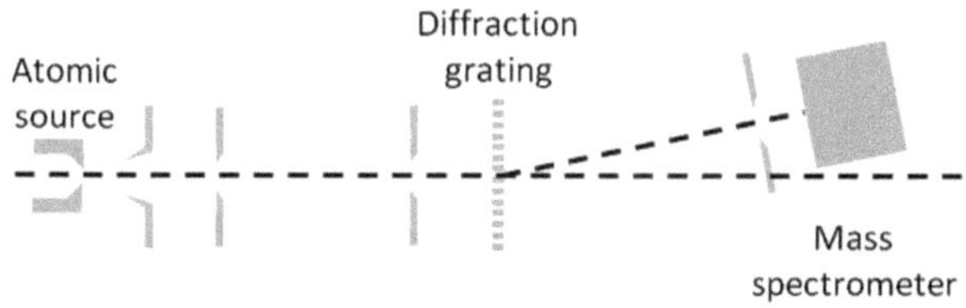

Fig. 2.14 An atomic source produces He atoms and He_2 dimers. The dimers are separated by diffraction and analyzed using a mass spectrometer

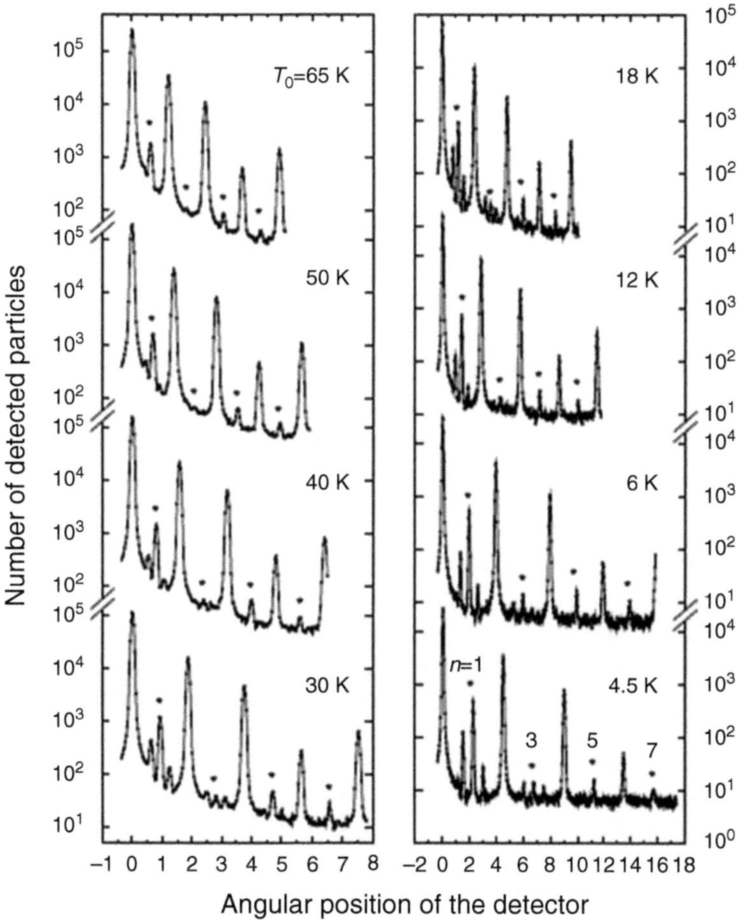

Fig. 2.15 Spectra of He atoms and dimers of He obtained after diffraction for several temperatures of the atomic source. The positions of the dimers are indicated by points

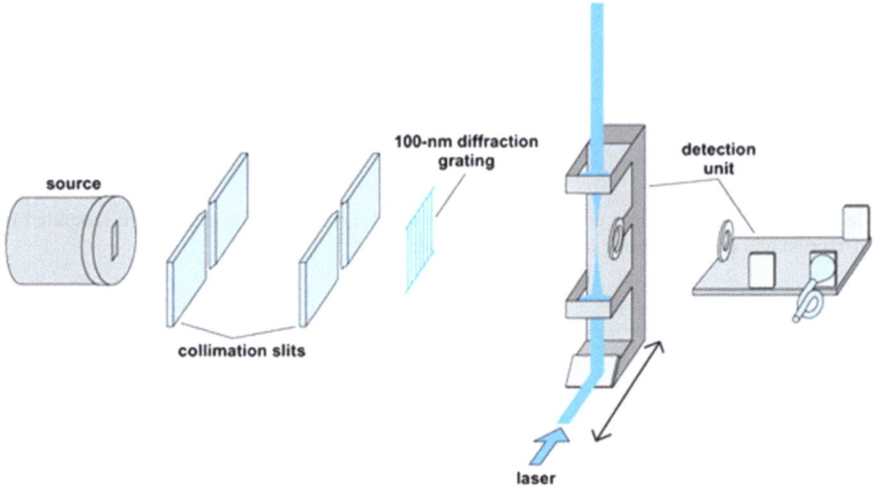

Fig. 2.16 Experimental set-up for C_{60} interferences. The source produces the fullerenes that are deviated by a grating. A laser induces a phase and the products are finally detected

2.2.4 Interferences with Large Molecules and Clusters

We know that the wave nature of particles is increasingly difficult to observe as their mass increases, because the de Broglie wavelength decreases strongly. However, 10 years ago, the wave nature of fullerenes was demonstrated [15, 16]. To observe the interference pattern, the experimental set-up was similar to that used for atoms [15]. The fullerenes were heated, and the velocity selected (Fig. 2.16). The most probable velocity was of the order of 200 m/s, inducing a wavelength of ~3 pm. The fullerenes were then passed through a grating, consisting of slits of a width of 50 nm and separated by a distance of 100 nm. The fringe spacing was thus ~34 μm. The diffracted atoms were then ionized using a laser, and the resulting ions accelerated and directed towards a conversion electrode.

Figure 2.17 presents the typical results for the intensity. On the bottom, the profile beam is given without grating. With the grating (top of the figure), the diffraction pattern is observed (open circles). The essential features of the interference pattern can be understood using standard Kirchhoff diffraction theory for a grating with a period of 100 nm, taking into account both the finite width of the collimation and the experimentally determined velocity distribution. The parameters in the fit were the width of the collimation, the gap width s_0 of a single slit opening, the effective beam width of the detection laser, and an overall scaling factor. This model, which assumes all grating slits to be perfect and identical, reproduces the central peak of the interference pattern very well, as shown in Fig. 2.17a.

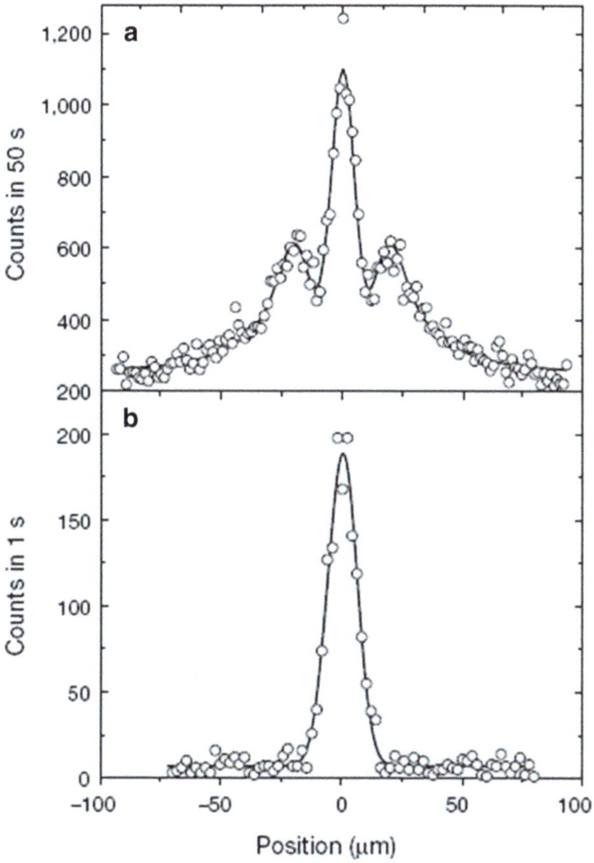

Fig. 2.17 Interference pattern produced by C_{60} molecules. (**a**) Experimental recording (*open circles*) and fit using Kirchhoff diffraction theory (*continuous line*). The expected zeroth- and first-order maxima can be clearly seen. (**b**) The molecular beam profile without the grating in the path of the molecules

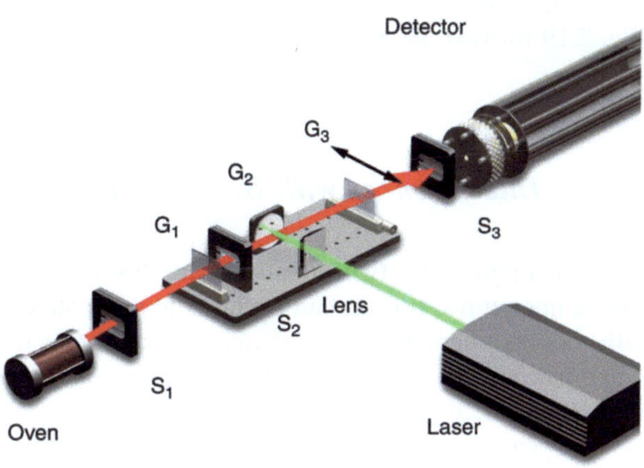

Fig. 2.18 Experimental device devoted to interference observation for large organic molecules [17]

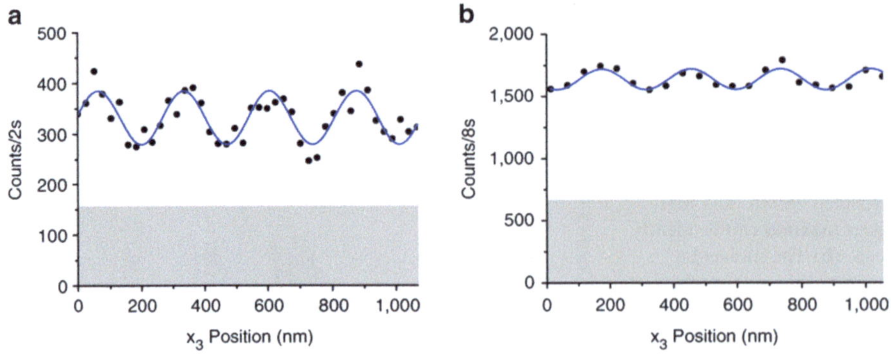

Fig. 2.19 Quantum interference is observed for two species, as a function of the third grating position. (**a**) perfluoroalkylated nanospheres $C_{60}(C_{12}F_{25})_8$. (**b**) $C_{60}(C_{12}F_{25})_{10}$. The *shaded area* indicates the detector dark rate

The wave nature of very massive organic molecules was demonstrated in 2011 [17]. An effusive source emitted molecules of e.g., $C_{60}(C_{12}F_{25})_8$ and $C_{44}H_{30}N_4$ whose masses were greater than 600 amu. The velocity of the molecules was selected through a sequence of three slits (Fig. 2.18).

Each slit of G_1 imposed a constraint onto the transverse molecular position that led to a momentum uncertainty. The latter turned into a growing delocalization and transverse coherence of the matter wave with increasing distance from G_1. The second grating G_2 was a laser light wave with a wavelength of $\lambda = 532$ nm. The interaction between the electric laser light field and the molecular optical polarizability created a sinusoidal potential, which phase-modulated the incident matter waves. The molecular nanostructure was sampled by scanning a third grating G_3 across the molecular beam while counting the number of the transmitted particles in a quadrupole mass spectrometer (QMS).

The interference fringes can be observed with a very good contrast, as shown in Fig. 2.19 for two species.

2.2.5 *Interferences with Bose-Einstein Condensates*

The indistinguishability of identical particles has important consequences. At very low temperatures, bosons can occupy the same state, giving rise to Bose-Einstein (BE) condensates. To obtain BE condensates, one method has been to employ a combination of laser cooling and evaporation cooling to produce dense, cold atomic clouds (Fig. 2.20).

Two years after the first creation of BE condensates, an experiment was performed to observe the interference pattern of two condensates of Na [18]. After letting the

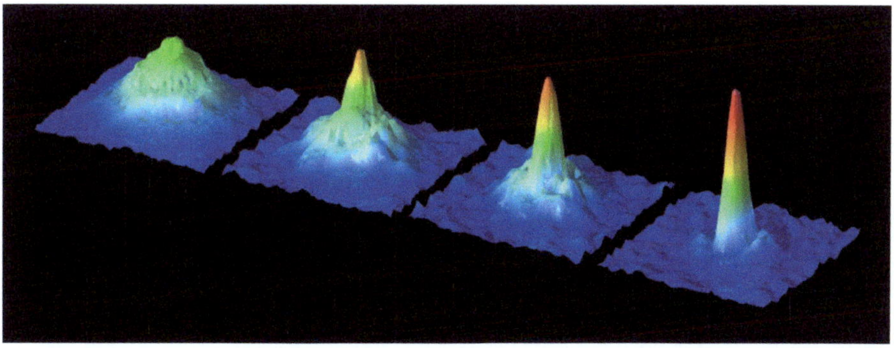

Fig. 2.20 Atoms of Rb (*left-hand side*) are cooled in a double magneto-optical trap. The atoms are laser-cooled to a temperature of ~0.03 K, and then transferred to a magneto-optical trap where they are cooled by radio-frequency evaporation. After ~5 s, a temperature of 500 nK is reached

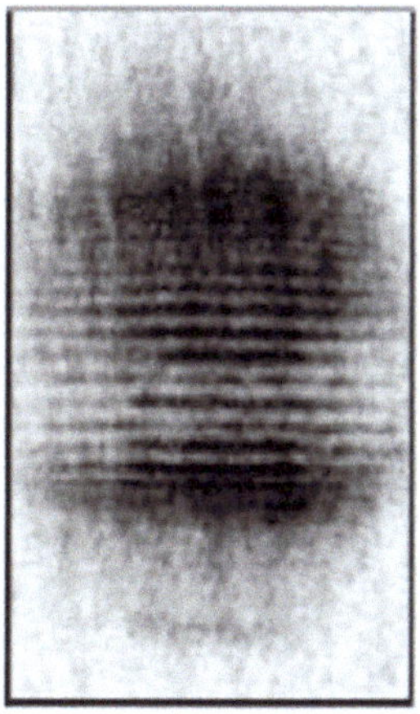

Fig. 2.21 Interference pattern of two expanding condensates observed after 40 ms time of flight [18]

condensates expand and overlap, interference between them was observed, with high-contrast fringes. The observation of interference (Fig. 2.21) demonstrated that BE condensates were coherent, i.e., they behaved like lasers.

References

1. Bohr, N.: On the constitution of atoms and molecules. Philos. Mag. **26**, 1 (1913)
2. de Broglie, L.: La physique quantique reste-t-elle indéterministe?, Revue d'histoire des sciences et de leurs applications, tome 5, no4 (1952)
3. de Broglie, L.: Recherche sur la théorie des quanta, Thèse, Paris (1924)

4. de Broglie, L.: *La* Mécanique ondulatoire et la structure atomique de la matière et du rayonnement. J. Phys. Paris, série VI, t. VIII 225, 15–31 (1927)
5. Davisson, C.J.: The discovery of electron waves. Nobel lecture 1937, Physics 1922–1941, Elsevier Publishing Company, Amsterdam (1965)
6. Elsasser, W.: Bemerkungen zur Quantenmechanik frier Elektronen. Naturwissenschafter **13**, 711 (1925)
7. Davisson, C.J., Kunsman, C.H.: Scattering of low speed electrons by platinum and magnesium. Phys. Rev. **22**, 242 (1923)
8. Fermi, E., Marshall, L.: Interference phenomena of slow neutrons. Phys. Rev. **75**, 666 (1947)
9. Rauch, H., Treimer, W., Bonse, U.: Test of a single crystal neutron interferometer. Phys. Lett. A **47**, 369 (1974)
10. Carnal, O., Mlynek, J.: Young's double-slit experiment with atoms: a simple atom interferometer. Phys. Rev. Lett. **66**, 2689 (1991)
11. Keith, D.W., et al.: An interferometer for atoms. Phys. Rev. Lett. **66**, 2693 (1991)
12. Miffre, A., et al.: Atom interferometry. Phys. Scr. **74**, C15 (2006)
13. Bordé, C.J.: Atomic interferometry with internal state labeling. Phys. Lett. A **140**, 10 (1989)
14. Baudon, J., Mathevet, R., Robert, J.: Atomic interferometry. J. Phys. B **32**, R173 (1999)
15. Arndt, M., et al.: Wave-particle duality of C_{60} molecules. Nature **401**, 680 (1999)
16. Arndt, M., et al.: High contrast interference with C_{60} and C_{70} molecules. C. R. Acad. Sci. Paris, t.2, Série 4 581 (2001)
17. Gerlich, S., et al.: Quantum interference of large organic molecules. Nat. Com. **2**, 263 (2011)
18. Andrews, M.R., et al.: Observation of interference between two Bose condensates. Science **173**, 637 (1997)

Chapter 3
Electron Interferences Using Macroscopic and Nanoscopic Interferometers

3.1 Electron Interferences with Macroscopic Interferometers

3.1.1 Interferences Using an Electron Beam: The Young Experiment Revisited

Since the pioneering work by Davisson and Germer in 1927, many experiments have been performed to validate the hypothesis of wave-particle duality. In 1961, a real double-slit experiment was conducted by C. Jönsson, from Tübingen University [1]. We need to mention this remarkable experiment, since it was an exact replica of Young double-slit experiment with light. The principle is as follows. A 50 kV electron beam crosses a system composed of one up to five slits, and the scattered electrons are detected on a film. This principle is simple, but the realization was not straightforward. The biggest difficulty was to make the slits, since the wavelength associated with electrons is much smaller (0.005 nm) than the atomic dimensions.

A thin layer of silver was evaporated on a glass plate (Fig. 3.1), so that the plate became electrically conductive. Strips of polymerized hydrocarbon were printed on the silver layer. The plate was then coated with a copper layer in an electrolytic bath. The portions of the silver layer covered by the polymerized material remained uncoated. The copper foil could then be peeled from the glass carrier, leaving behind the silver layers and the polymerized strips at the positions of the slits. The length of the slits was ~50 μm. The slits had a width of ~0.5 μm and a spacing of ~2 μm.

To satisfy coherent conditions, the scattering angle of electrons had to be very small. The diffraction pattern was expected to be so small that it was considerably enlarged by introducing electrostatic lenses. Using a camera to record interference pattern, the exposure time varied from 20 s to 3 mn.

F. Frémont, *Young-Type Interferences with Electrons*, Springer Series on Atomic, Optical, and Plasma Physics 77, DOI 10.1007/978-3-642-38479-0_3,

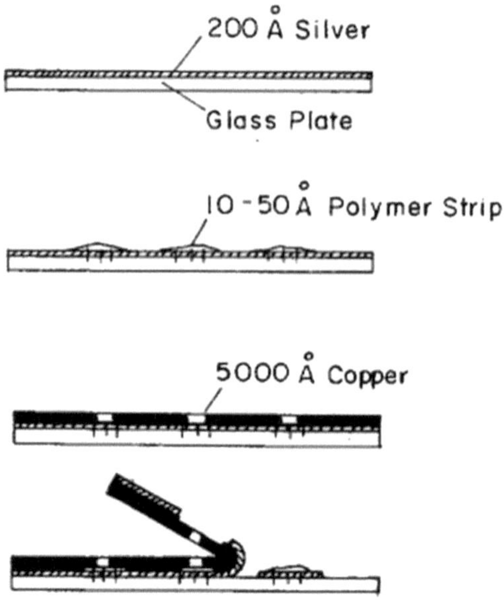

Fig. 3.1 Production of slits in Jönsson's interference experiment [1] (see text)

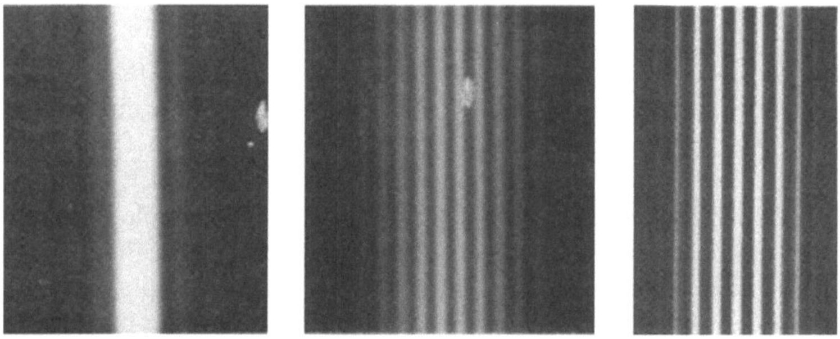

Fig. 3.2 Interference pattern observed after electrons scattered on one (*left-hand side*), two (*center*) and five (*right-hand side*) slits [1]

The result is presented in Fig. 3.2 for one, two and five slits. Using one slit (left-hand side of the figure), a diffraction figure is obtained, with a maximum at the center. With two and five slits, interference patterns appear. Maxima and minima of intensity are all the pushed closer as the number of slits increases.

3.1.2 *Feynman Thought Experiment*

In 1964, Richard Feynman, famous professor and researcher, published his lecture notes. In his book on quantum mechanics [2], he made a detailed description of an interference experiment with electrons. After a brief comparison with what would happen when bullets crossed two apertures, he showed the similarities between photon and electron double-slit experiments. In both cases, interference fringes are visible and fringe spacing can be calculated using the same formula.

Feynman conceived this experiment by himself, as a "thought experiment". He explicitly underlined: "We should say right away that you should not try to set up this experiment. This experiment has never been done in just this way. The trouble is that the apparatus would have to be made on an impossibly small scale to show the effects we are interested in. We are doing a "thought experiment", which we have chosen because it is easy to think about. We know the results that would be obtained because there are many experiments that have been done, in which the scale and the proportions have been chosen to show the effects we shall describe". This statement clearly shows that Feynman was not aware of the Jönsson experiment that had been performed two years earlier. One likely reason for this is that Jönsson's results were first published in German in a German journal.

It should be pointed out that an erroneous interpretation is frequently made of Feynman's description of interferences. In many publications, the Feynman thought experiment is reported as an experiment in which one single electron crosses the double-slit apparatus at a time. In fact, this point was never raised by the author. Instead, he described the detection of electrons as follows. Imagine a detector that emits a click each time a group of electrons reaches the detector. The clicks have the same intensity whatever the number of electrons in each. If the source delivers fewer electrons, let us say 100 electrons instead of one million, fewer electrons will be detected, but the noise of each click will be the same. To summarize, Feynman spoke about groups, or ensembles, that could contain one electron equally well as one million electrons. Thus, he did not say anything about what would happen when one electron at a time reached the detector. He only considered the situation where the electron beam was attenuated.

His conclusions are nevertheless essential for understanding the interference phenomenon. If we try to know which hole the electron passed through, then the interference pattern will be lost. The spatial distribution resembles that obtained with gun bullets (Fig. 3.3). On the contrary, if we don't try to detect where the electrons passed through, interferences are created and can be observed (Fig. 3.4.). In other words, Feynman concluded, in the first case the electrons pass through one hole or the other while, in the second case, we don't know through which hole they passed. The important point here is that Feynman does not say that the electrons cross both holes at the same time.

3.1.3 Interferences Using an Attenuated Electron Beam

About 10 years later, Italian researchers performed the first experiment with an attenuated electron beam [3]. The experiment consisted of the visualization of interference fringes after the electron beam scattered on a kind of slip of card, provided by an electrostatic Fresnel biprism (Fig. 3.5).

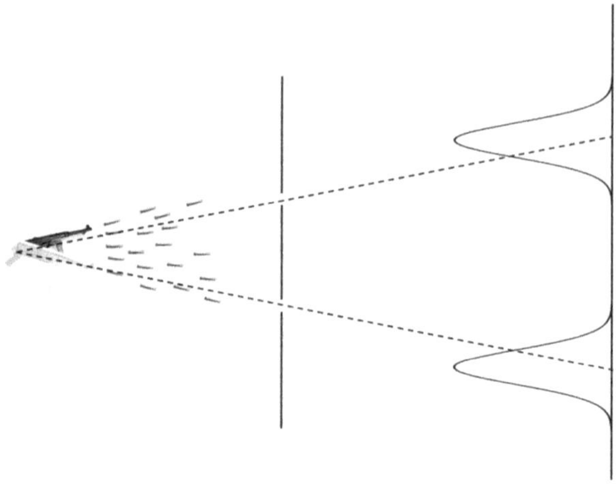

Fig. 3.3 Bullets are directed randomly toward two holes. The bullets that pass are counted on a screen. They are distributed on both sides on axes defined by the gun direction and the holes (*dashed lines*)

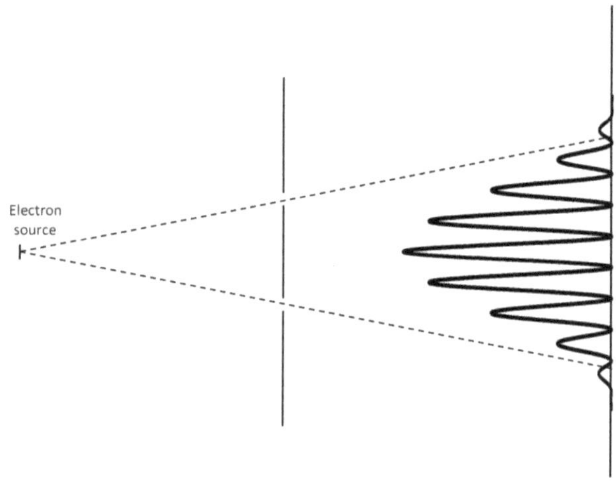

Fig. 3.4 A source produces electrons that scatter through two holes. In contrast to what happens with bullets, interferences are visible, due to the fact that electrons behave like a wave, so that the amplitudes are coherently added

The electron beam is provided by a source S, and the electrons passing on each side of a thread (equivalent to Young's slip of card), are focused by means of an electric field. The scattered electrons interfere in the hatched region on the figure and are detected on a television screen or a photographic plate. Due to the weakness of fringe spacing, two electrostatic lenses are added to increase the visibility by a factor of 100.

As claimed by the authors, the most important point in this experiment is the statistical aspect of the fringe formation. The electrons were detected over a total time of 0.1 s, for different intensities of the primary beam (Fig. 3.6). On Fig. 3.6a, b, the beam intensity is sufficiently weak that individual plots are observed, due to individual impacts. When the intensity increases, the fringes become increasingly visible, showing the wave behavior of electrons. The electrons reach the screen randomly, but coherently. "The complete interference pattern registered on the

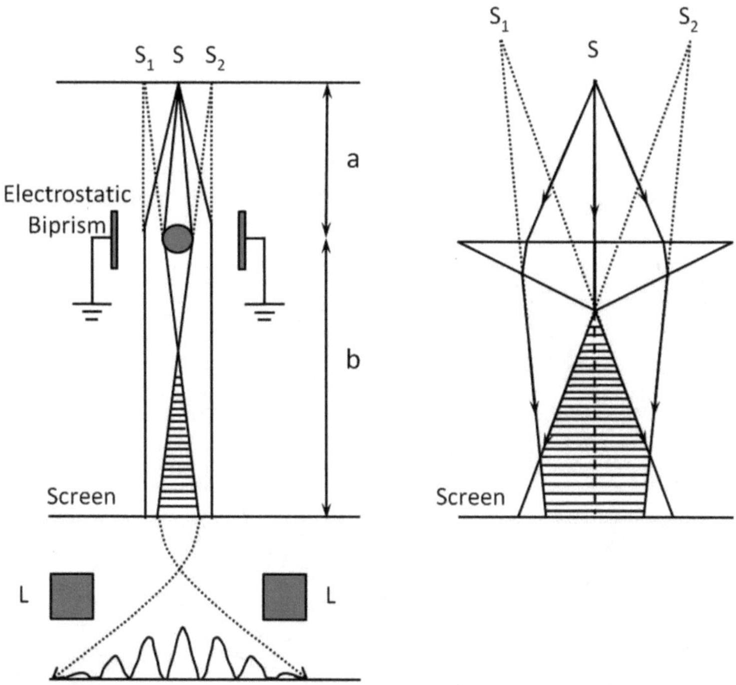

Fig. 3.5 Electrostatic biprism for electrons (*left-hand side*) and Fresnel biprism (*right-hand side*) for photons. After deflection, the source provides two virtual sources S_1 and S_2. The scattered particles overlap (*dashed region*), and interferences are created (*hatched regions*). To observe the fringes in the case of electrons, lenses *L* are added to increase the visibility

photographic plate is the sum of individual events, each due to the interaction between a single electron and the interference apparatus" [3]. The authors showed that it is equivalent to fix the current density in order to detect each electron every 0.04 s and to register all the events over a few minutes.

In 1989, Japanese researchers performed a similar experiment [4], using an analogous experimental set-up. Surprisingly, the authors said "the present experiment aim[s]ed at realizing the two-slit thought experiment for electrons", as though they were the first to do this. Moreover, they only cited the film made by the Italian group, and not the paper they published in 1976.

In 2003, each of these groups criticized the other over the authorship of the demonstration [5]. The Japanese group claimed that their experiments differed from the Italian group's experiments in two following respects:

- Their experiments were carried out from beginning to end with constant and extremely low intensities.
- The formation of fringes could be observed as a time series, the electrons were accumulated over time to gradually form an interference pattern.

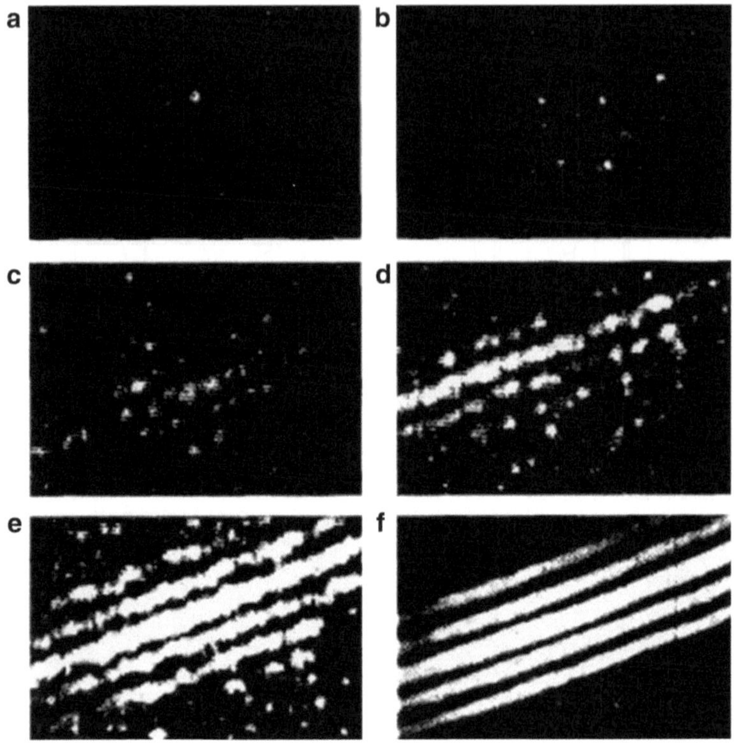

Fig. 3.6 Electron interference fringe pattern filmed at increasing current densities, and shown on a television screen. At low densities, (**a** and **b**), individual plots are observed. As density increases, interference fringes appear

The Italian group answered these criticisms: “Obviously the electron-detection system used by the Hitachi group in 1989 was more sophisticated than the one we used in 1974. However, the sentence on page 118 of the paper by Tonomura et al., which states that in our film we ‘showed the electron arrival in each frame without recording the cumulative arrivals’, is not correct: this can be seen by watching the film and looking at figure 1 of our 1976 paper.”

Two remarks need to be made on the work of these groups. First, both experiments are important because they confirm the wave behavior of electrons by mimicking Feynman’s thought experiment, even though the two slits are replaced by an electrostatic equivalent of the slip of card. Second, and most importantly, following Grangier and collaborators, these experiments are not true single electron experiments because they were performed using an attenuated beam. The single electron condition is not fulfilled; one can just say that there is one electron at a time, on average. It was not until 2004 that a single-electron interference experiment was theoretically described (cf. Chap. 4).

3.2 Electron Interferences with Nanoscopic Interferometers

3.2.1 Fundamental Interest of the Approach

With nanoscopic interferometers, such as atoms or molecules, the mechanisms leading to interferences are very different from those occurring with macroscopic interferometers. When an electron crosses two slits, the interferences are produced by short-range interactions. However, an electron experiences an atom or a molecule field during an infinite time due to Coulomb-type interaction, which is a long-range interaction. Thus the theoretical approach to describing interferences at the atomic scale is expected to be a real challenge, because it is necessary to study what happens in space, and also in time.

In a collision between an electron and an atom, the processes are not unique. The electron may scatter on the atom, but may also excite or ionize the atom. Due to the multiplicity of processes that can lead to the same state for the electron after the collision, interferences can have several origins, and thus can reduce the effect we would expect.

Interference study is a good test for existing theories. Depending on the degree of sophistication, we would expect theoretical results to describe the experimental results to a greater or lesser extent. In fact, at present, due to the complexity of the problem, many approximations need to be made that distort reality.

Finally, interferences may give information not only on the wave nature of electrons, but also on the processes themselves that occur during the collisions.

One should also mention difficulties that occur with the electron source. Electrons can be supplied directly from a filament. However, they can also be produced indirectly, for example by ionization or the Auger effect (cf. Chap. 4). Therefore, instead of an electron scattering on an atom, an electron from this first atom can end up scattering on a second atom. The initial two-body problem then becomes a three-body problem.

Before entering into the details of electron interferences at the atomic scale, it is thus necessary to briefly describe the processes that lead to interference phenomena. These processes will be introduced in two parts: collisional processes that occur when the collision partners are close to each other (a few nm), and post-collisional processes, where the collision partners are far away from each other (typically 100 nm or more).

3.2.2 *Processes in Atomic Collisions*

3.2.2.1 Collisional Processes in Atomic Collisions

Suppose as a simplified example, a fully stripped ion A^{Z+} colliding with a target (atom or molecule) B¸ which is at rest before the collision. The incident ion has a velocity $\overrightarrow{v_P}$ and a corresponding kinetic energy E_p. The relative importance of the processes that occur during the collision mainly depends on the ratio between v_P and the velocity v_e of the electron that will be involved during the collision (this electron is called active electron). Three velocity regions are distinguished:

If $v_P \gg v_e$, the dominant processes are excitation and/or ionization of the target. In the case of excitation, the projectile gives part of its energy to the active electrons that are excited on a given level (Fig. 3.7).

There is also a probability for the electron to be excited into the continuum (Fig. 3.8). In this case, the electron is no longer bound to the target. The target is ionized, and the electron has a certain kinetic energy.

If $v_P \ll v_e$, the projectile spends enough time in the vicinity of the target to capture one or several electrons from it (Fig. 3.9). This process is called charge transfer or electron capture. In most of the collisions, the projectile will be in an excited state after the collision.

If v_P is of the same order of magnitude as v_e, ionization, excitation and electron capture compete and these different potential outcomes of the collision have equivalent probabilities.

Finally, there is a probability that the projectile does not influence the initial electronic state of the target. The projectile only gives energy and momentum to the target (elastic scattering).

The different processes can be summarized as follows:

Excitation[a]	$A^{Z+} + B \rightarrow A^{Z+} + B^{**}$
Ionization[b]	$A^{Z+} + B \rightarrow A^{Z+} + B^{q+}$
Electron capture[c]	$A^{Z+} + B \rightarrow A^{(Z-r)+} + B^{r+}$
Elastic scattering	$A^{Z+} + B \rightarrow A^{Z+} + B$

[a]The asterisks means that B is in a doubly excited state
[b]q electrons are removed from B
[c]r electrons are captured by the projectile

3.2.2.2 Post-collisional Processes in Atomic Collisions

To illustrate a post-collisional process, suppose a collision between a He^{2+} ion and a molecular target H_2. At low projectile velocity, He^{2+} may capture both target electrons in a doubly excited state, $2\ell 3\ell'$ for example. The projectile may become deexcited via a radiative cascade, as shown in Fig. 3.10. The 2ℓ and $3\ell'$ electrons reach the lower levels independently, emitting each a photon.

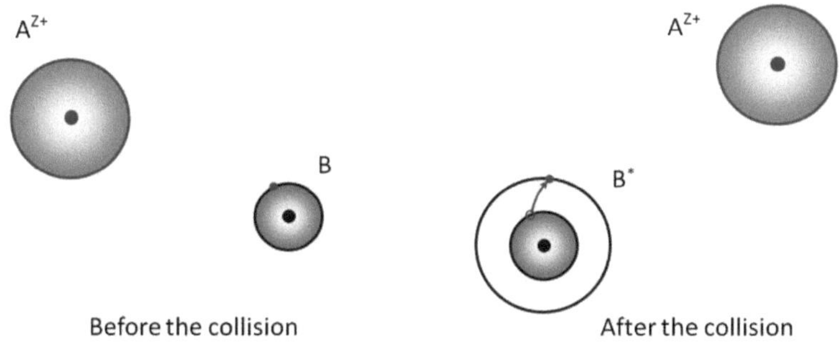

Fig. 3.7 Process of excitation in an atomic collision. The active electron of B is excited on a higher level (*right-hand side of the Figure*)

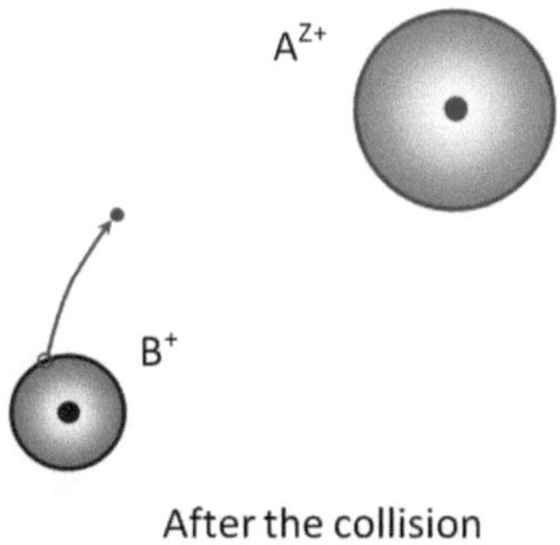

Fig. 3.8 In the ionization process, one electron is removed from the target with a certain kinetic energy

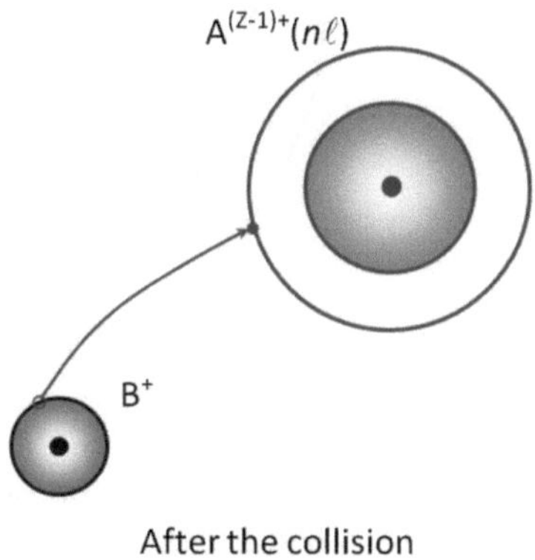

Fig. 3.9 The projectile captures one electron and is in an excited state defined by the principal quantum number n and the angular momentum ℓ

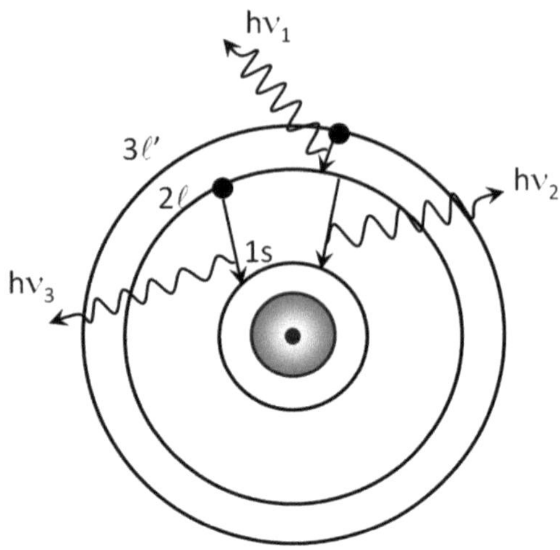

Fig. 3.10 Radiative cascade after a capture of two electrons in $2\ell 3\ell'$ configuration

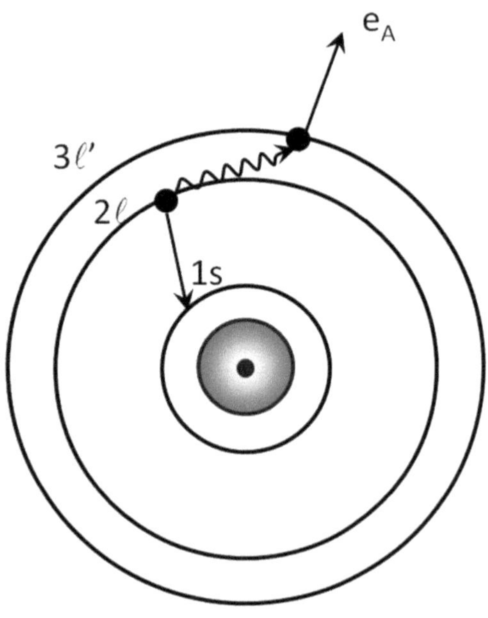

Fig. 3.11 Emission of an Auger electron e_A after capture of two electrons in $2\ell 3\ell'$ configuration

There is also a non-negligible probability for the 2ℓ electron to reach the 1s orbital, giving its energy to the $3\ell'$ electron that will be ejected with a well defined energy (Fig. 3.11). This electron is called Auger electron.

Finally, because the target is a molecule, the ejection of the Auger electrons is accompanied by the fragmentation of the residual target. For example, in the case of a double ionization of H_2 molecule, the residual target H_2^{2+} separates into two protons that have, at infinity, a velocity of ~0.0195 a.u., corresponding to a kinetic energy of ~9.5 eV for each proton.

3.2.3 *Approximations in the Collision and Post-collision Calculations*

3.2.3.1 Separation of the Collision and Post-collision Processes

In most of the collisions that have been studied up to now, the projectile energies are larger than 10 keV, corresponding to velocities larger than ~0.1 a.u. The collision interaction is effective on distances of ~20 a.u. or less, corresponding to the overlap of the classical orbitals of the target and projectile. Consequently, the interaction time t_c is less than 200 a.u. (~5 $\times$ 10^{-15} s). From another perspective, the Auger and radiative effects have typical deexcitation times t_A and t_R of ~10^{-13} s and 10^{-10}s, respectively. These times are much larger than t_c, thus we can separate collisional processes from post-collisional processes. If the projectile velocity decreases, for example by a factor of 100, this approximation is no longer valid.

This approximation is, nonetheless, useful for a theoretical treatment of the collision. In the case of a He^{2+} + H_2 collision, for example, the problem is a five-body problem (He, two protons and two electrons). Within the approximation, however, the capture and post-collisional processes can be approximated as three-body problems (He, H and one electron).

3.2.3.2 Projectile Motion

The electrons can be treated using quantum mechanics. Nuclei, in contrast, are classically treated, since the associated wavelength is very small compared with other distances. For example, an energy of 100 keV for an ion corresponds to a wavelength of ~10^{-11} m, which is 1,000 times shorter than the collision length. However, this approximation is again no longer valid when the projectile velocity decreases, since the wavelength then increases.[1]

Another approximation is also valid at relatively high projectile velocities ($v_p > 0.1$ a.u.). At such velocities, the projectile trajectory can be considered as a straight line. The scattering angle θ is inversely proportional to the incident energy E_p, and an energy of 100 keV induces an angle of a few thousandths of a radian. Furthermore, in this case, the velocity is supposed to be constant during the collision. Within this approximation, it is useful to introduce the impact parameter b, defined as (Fig. 3.12)

$$\overrightarrow{R} = \overrightarrow{b} + \overrightarrow{v_p}\, t \tag{3.1}$$

where R is the distance between the nuclei (internuclear distance).

[1] The correspondence between atomic units (a.u.) and usual units are explained in Appendix 1

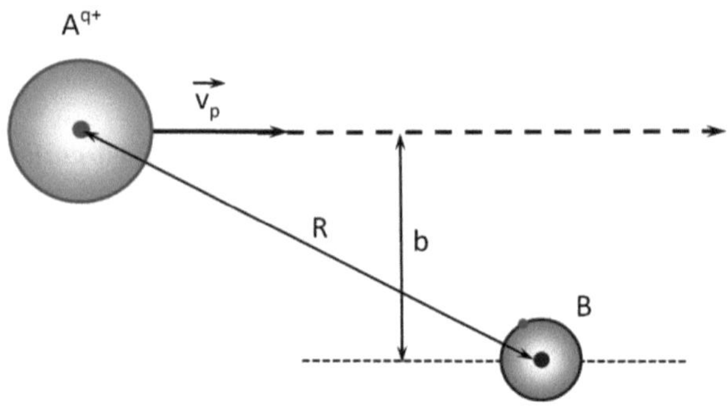

Fig. 3.12 The projectile A^{q+} scatters on B with a high velocity v_p. The trajectory is a *straight line*, and v_p is constant

Again, these two latter approximations cease to be valid as v_p decreases. At low velocities, the projectile trajectory is no longer a straight line, and the projectile velocity varies throughout the collision (Fig. 3.13).

3.2.3.3 Electronic Basis

In its initial state, the electron is in the fundamental state. During the collision, the electron is excited, ionized and/or captured. In principle, the final state is therefore an infinite combination of these individual states, each state being the consequence of the processes that occur during the collision. Consequently, the basis is composed of an infinite quantity of states. In practice, the problem as it stands is insolvable. In fact, in most collisions, only a few states are considered, because the probability of these outcomes is much higher than the others. Thus, it is sufficient to introduce a truncated basis, formed by the most probable states.

3.3 The Role of Electrons in Ion Interferences

3.3.1 Elastic Process in Asymmetric and Symmetric Collisions

Before the advent of Electron Cyclotron Resonance (ECR) sources, the collisions were investigated at high and low velocities with atomic projectiles or monocharged ions, where elastic scattering may play a significant role. Scattering cross sections were systematically measured as a function of the scattering angle θ, i.e., the angle between the initial and final direction of the projectile, for asymmetric collisions $A^+ + B$ as well as symmetric collisions $A^+ + A$, at collision energies smaller than ~1 keV.

We will first focus on asymmetric collisions. Figure 3.14 shows typical differential cross sections in the case of a $He^+ + Ne$ collision [6], investigated at energies

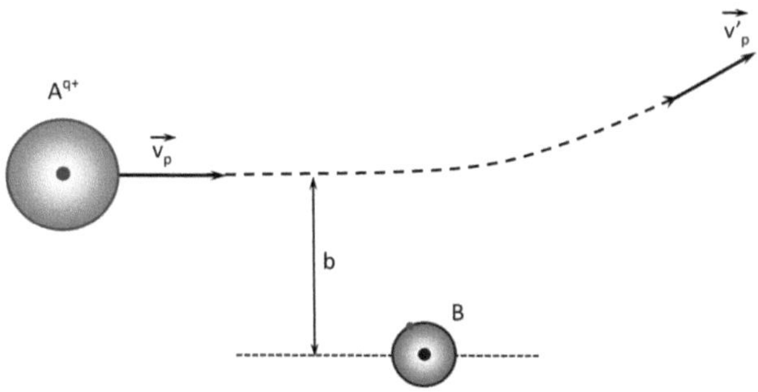

Fig. 3.13 The same elements are shown as in Fig. 3.12, but at low velocity. The trajectory is *curved*, and the final velocity is different from the initial velocity

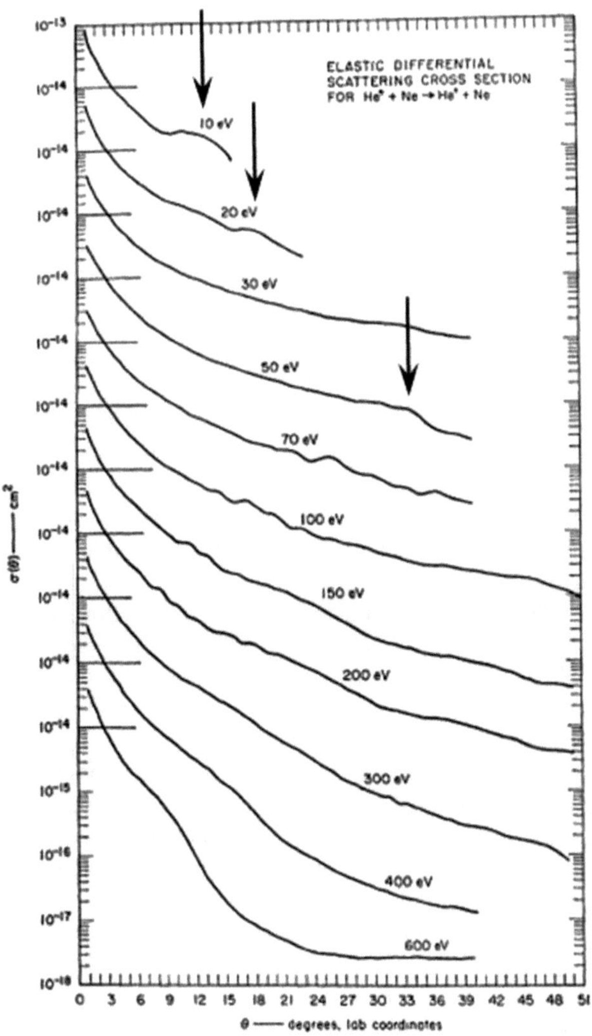

Fig. 3.14 Scattering differential cross sections for He^{+} + Ne collisions, at energies from 10 eV up to 600 eV. The *arrows* show the structures originating from the potential variation due to the presence of electrons

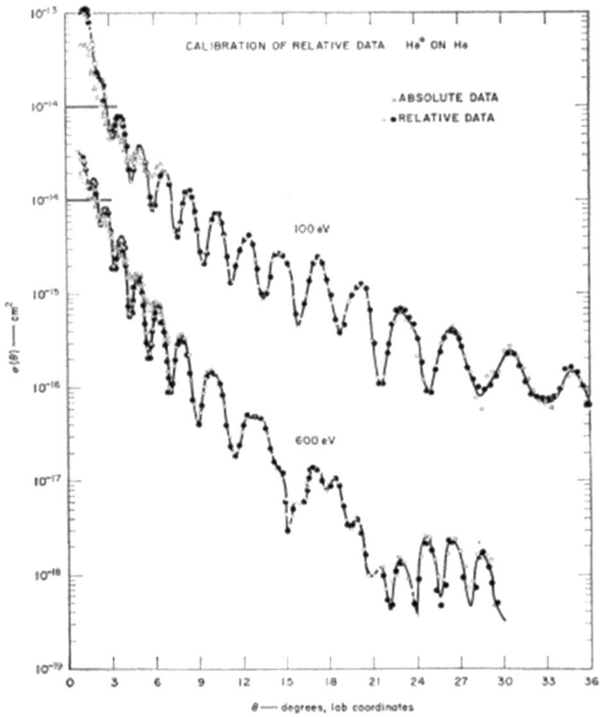

Fig. 3.15 Scattering differential cross sections for He^{+} + He collisions, at energies of 100 eV and 600 eV. Well defined oscillations are visible, which are due to the indiscernability of the collision partners

ranging from 10 eV (top of the Figure) to 600 eV (bottom of the figure). In the absence of electrons, the cross section would follow the well-known Rutherford law, and would decrease as $1/\sin^4\theta$. Due to the presence of the electrons on the projectile and on the target, perturbations caused by distorted potentials appear, giving rise to structures that are clearly visible at the largest angles (indicated by arrows on the figure).

The angular dependence of the cross section is completely different in the case of symmetric collisions. We will first analyze the simplest case of He^{+} + He collision, whose scattering differential cross sections are presented on Fig. 3.15. In addition to monotonous decreasing variation of the cross section, with high amplitude oscillations are observed.

In fact, when looking more carefully the cross sections, two kinds of oscillations can be distinguished. At scattering angles smaller than 15°, the period of the oscillations increases as the angle increases. At 600 eV, at angles larger than 15°, oscillations with small periods appear for the two main oscillations. This means that the interferences have two origins.

The interpretation of these oscillations is as follows: The main oscillations originate from orbital overlap. In the case of the collision He^{+} + He for example, the active electron can be on the projectile ($1s_P$ orbital) or on the target ($1s_T$ orbital). The molecular orbital corresponding to the electron state of the molecule (He^{+}–He) formed during the collision is a linear combination of both atomic orbitals $1s_P$ and $1s_T$. In fact, for this combination there exist two possibilities, either ($1s_P$ + $1s_T$) or

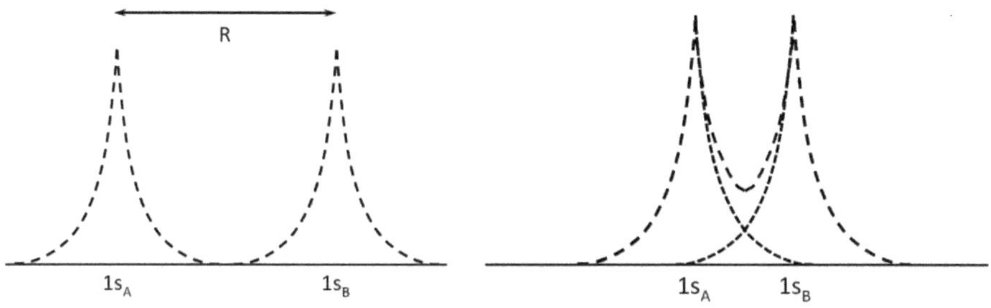

Fig. 3.16 Bonding orbitals. On the *left-hand side*, the nuclei are far from each other. *On the right-hand side*, a molecule is formed, since the orbitals overlap

($1s_P - 1s_T$), corresponding to bonding (Fig. 3.16) and antibonding (Fig. 3.17) orbitals, respectively. In the case of a bonding orbital, a molecule is formed at a certain distance (dashed curve on Fig. 3.16), as shown in Fig. 3.18 which shows potential energies of the system as a function of the internuclear distance. For antibonding orbitals, the curve is repulsive (full curve on Fig. 3.18).

The potential energies associating with these two possibilities are shown on Fig. 3.18. When the wave functions overlap, the potential energy decreases and reaches a minimum value at a distance R_o, then increases at internuclear distances smaller than R_o.

These potential energies are associated with classical trajectories, which are represented in Fig. 3.19. When the orbitals are bound, the projectile is first attracted to the target, due to the presence of the electron (full curve). Then, the projectile is repulsed due to the Coulomb potential between the two nuclei. On the contrary, the repulsive potential leads to a repulsive trajectory (dashed curve). As seen on Fig. 3.19, both trajectories may lead to the same final state, with the same scattering angle. Thus, because the trajectories are indistinguishable, interferences occur, leading to the oscillations seen in Fig. 3.15.

The oscillations superimposed onto the main oscillations at angles larger than 15° originate from the indiscernibility of the nuclei themselves. Thus, when a nucleus is detected, one cannot say if it is the projectile or the target (Fig. 3.20). This indiscernibility also leads to interferences.

3.3.2 *Resonant Electron Capture in Symmetric Collisions*

3.3.2.1 Experimental Observations

In the last paragraph, we described the role of a passive electron in the interference phenomenon. Here, the electron becomes active, since it is captured by the projectile. At the beginning of the 1960s, electron capture was investigated in symmetric collisions, such as $H^+ + H$, $H^+ + H_2$ [7] or $He^+ + He$ [8], involving low or fast charged ions and atoms.

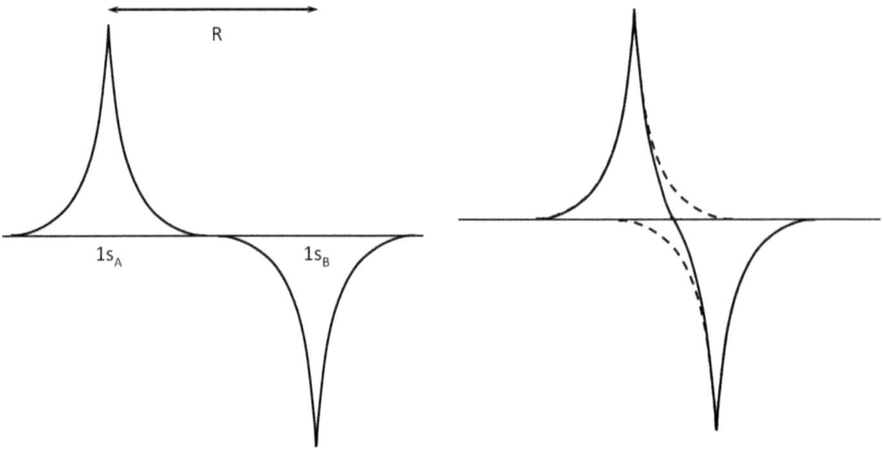

Fig. 3.17 Antibonding orbitals. On the *left-hand side*, the nuclei are far from each other. On the *right-hand side*, there is no overlap, so that the nuclei repulse each other

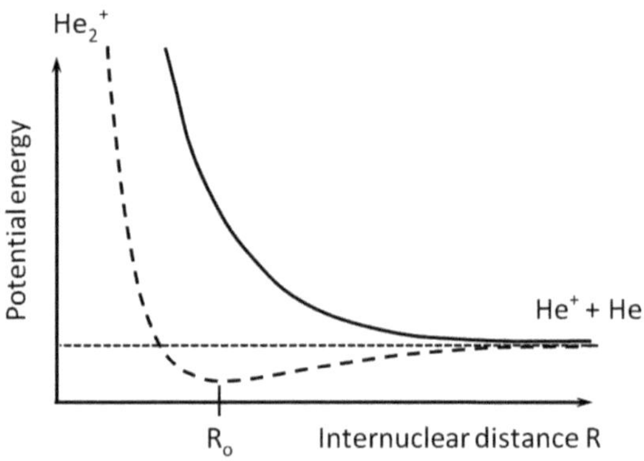

Fig. 3.18 Schematic potential energies for the collision system He^+ + He. At infinity, the nuclei are separated. At small internuclear distances, the orbitals can be repulsive (*full curve*) or attractive and repulsive (*dashed curve*)

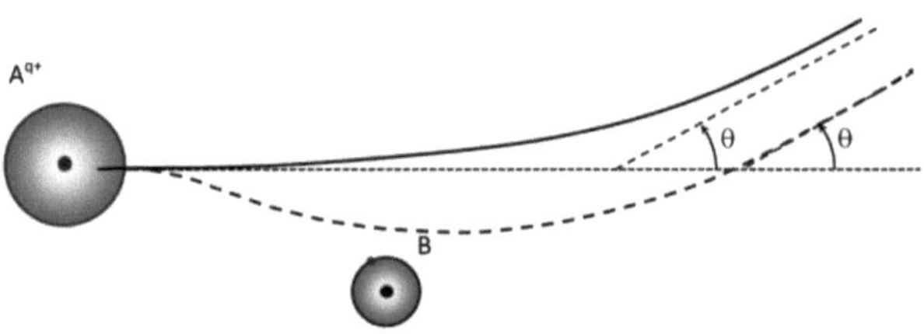

Fig. 3.19 Two different trajectories leading to the same scattering angle

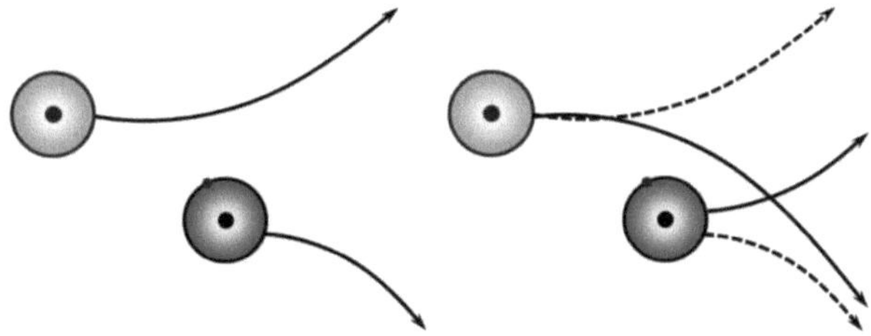

Fig. 3.20 On the *left*, the projectile is deflected upwards, while the target is deflected downwards. On the *right*, the trajectories are inverted

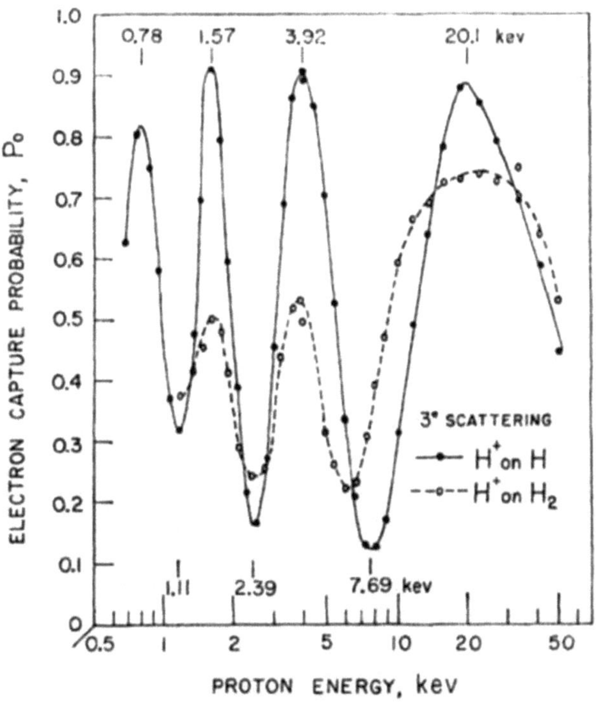

Fig. 3.21 Electron capture probability as a function of the projectile energy, for the collision systems H^+ + H and H^+ + H_2 [7]

For these systems, the authors discovered a remarkable experimental result at a large scattering angle, i.e., at a small impact parameter. When, plotted versus the projectile energy E_p, the electron capture probability showed several pronounced peaks (Fig. 3.21), for both H^+ + H and H^+ + H_2 collisions. In addition, the minima and the maxima were located at the same projectile energies, except at the largest, where a shift was observed. Finally, the molecular data showed peaks that were less pronounced than those in the atomic case.

When the H^+ + H capture probability is plotted as a function of the reciprocal velocity $1/v_p$, the oscillations are found to be equally spaced (Fig. 3.22).

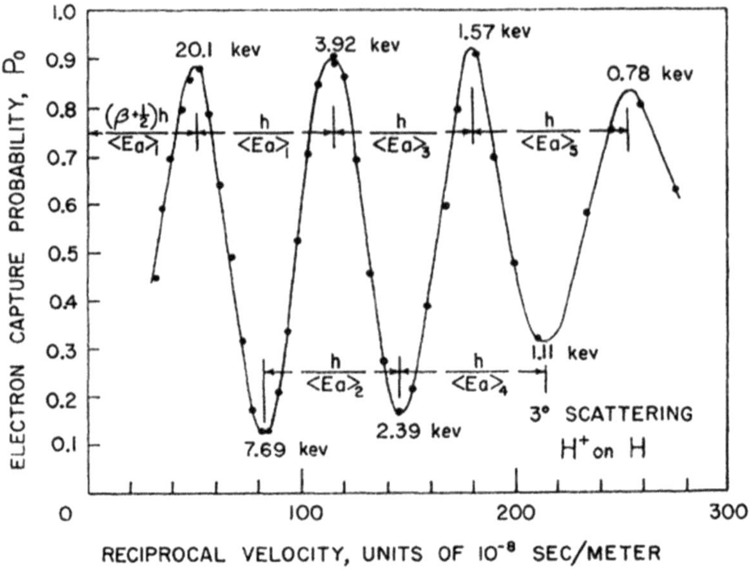

Fig. 3.22 Electron capture probability as a function of the reciprocal velocity $1/v_p$, for the collision systems $H^+ + H$ and $H^+ + H_2$ [7]

3.3.2.2 Interpretation

In symmetric collisions $A^+ + B$, the capture of one target electron leads to the final system $A + B^+$. In this case, the potential created by A^+ is identical to that created by B^+. Due to the indiscernibility of the potentials, the electron wave functions ψ_T and ψ_P on the target and the projectile, respectively, combine to give two asymptotic wave functions:

$$\psi_u = \frac{1}{\sqrt{2}}(\psi_T + \psi_P)$$
$$\psi_g = \frac{1}{\sqrt{2}}(\psi_T - \psi_P) \tag{3.2}$$

In addition, we assume that the motion of the projectile is sufficiently slow to consider that, at a fixed time, the state of the system is stationary. Thus, the dependence in time of the wave function is

$$\psi(t) = \psi_o e^{-iEt} \tag{3.3}$$

At a time $t + dt$,

$$\psi(t + dt) = \psi(t)\ e^{-i\frac{dE}{dt}tdt} \tag{3.4}$$

The phase shift during the collision is thus

$$\varphi(t) = \int_{-\infty}^{+\infty} \frac{dE}{dt} t dt \tag{3.5}$$

At infinity, when the collision is finished, the final wave function can be written

$$\psi_A(+\infty) \approx \underbrace{\frac{1}{2}\left(e^{i\varphi_u} + e^{i\varphi_g}\right)}_{capture} \psi_A(-\infty) + \underbrace{\frac{1}{2}\left(e^{i\varphi_u} - e^{i\varphi_g}\right)}_{\substack{elastic \\ scattering}} \psi_B(-\infty) \tag{3.6}$$

Finally, the capture probability is the square of the $\psi_B(-\infty)$:

$$P_{AB} \approx \sin^2\left(\frac{\varphi_u - \varphi_g}{2}\right) \tag{3.7}$$

with $\varphi_u - \varphi_g = \int_{-\infty}^{+\infty} \frac{d(E_u - E_g)}{dt} t dt$.

In the case where dE/dt is approximately constant and the potential is much smaller than E_p, expression (3.7) can be simplified

$$P_{AB} \approx \sin^2\left(\frac{1}{v_P} \int_{-\infty}^{+\infty} \frac{(E_u - E_g)}{\sqrt{R^2 - R_o^2}} R dR\right) \tag{3.8}$$

where R_o is the distance of closest approach.

The relation (3.8) clearly demonstrates the dependence of the probability on $1/v_p$. Note that if the electron is not captured, we find the case of the elastic scattering ($A^+ + B \rightarrow A^+ + B$) that was described in Sect. 3.2.2. The scattering probability is obtained by replacing the sinus function by a cosinus function in expression (3.8).

3.3.3 Electron Capture in Asymmetric Collisions

As a simplified example, we will consider a collision between a fully stripped ion A^{Z+} and H (atomic hydrogen). Since electron capture mainly takes place at internuclear distances of ~4–10 a.u., asymptotic potential curves are useful to describe the collision. Before the collision (entrance channel), the total electron energy is the energy of the electron on H, i.e., $E_i = -0.5$ a.u. When the electron is captured, the system is $A^{(Z-1)+}$ and H^+ (outgoing channel) and the total energy can be written $E_f = E(n) - Z/R$, where $E(n)$ is the energy of the bound electron on the projectile orbital characterized by the principal quantum number n (Fig. 3.23). At a certain distance, R_C, the two curves cross; this is the distance at which capture mainly occurs.

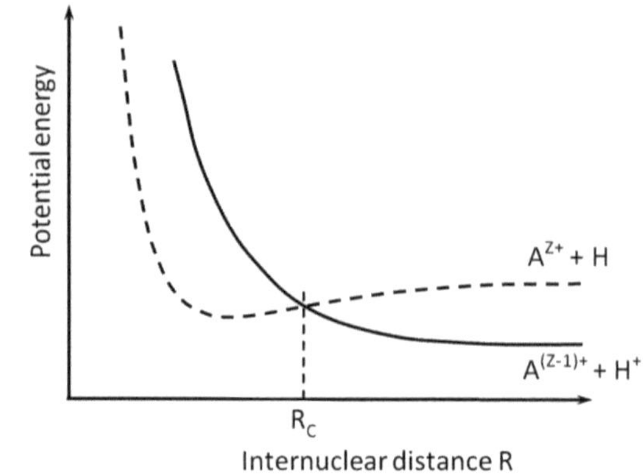

Fig. 3.23 Schematic potential curves, showing the crossing between the entrance channel (*dashed curve*) and the outgoing channel (*full curve*)

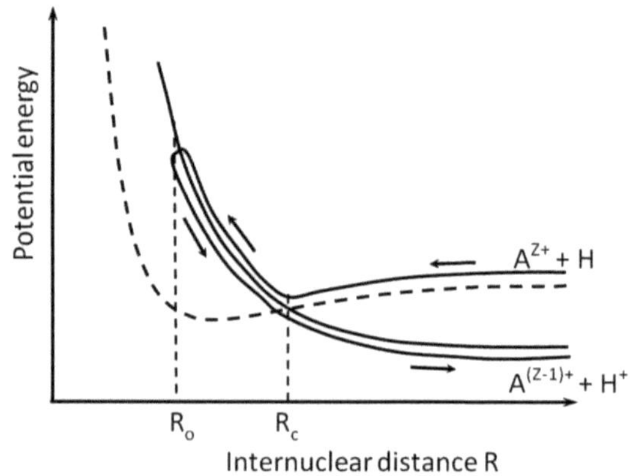

Fig. 3.24 At R_c, a first transition occurs between the entrance channel and the outgoing channel on the 'way-in' to the collision. On the 'way-our', the electron remains on the projectile

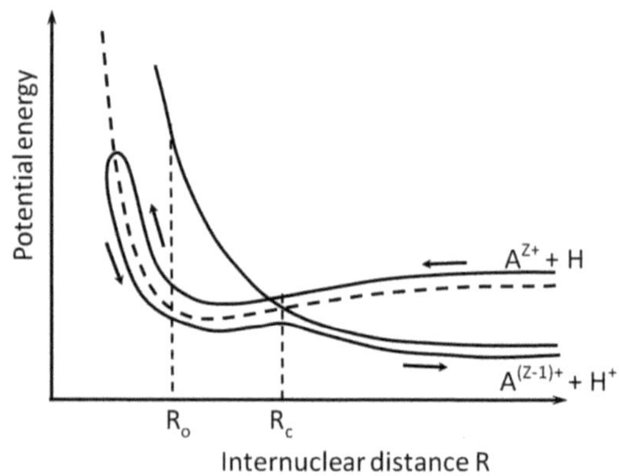

Fig. 3.25 In contrast with the previous figure, the electron is captured on the 'way-out' of the collision

There are two possibilities for the electron to be captured, which are depicted in Figs. 3.24 and 3.25. First, a transition may occur on the "way-in" of the collision, i.e., before reaching the distance of closest approach R_o (Fig. 3.24). Then, after being captured by the projectile, the electron remains upon it.

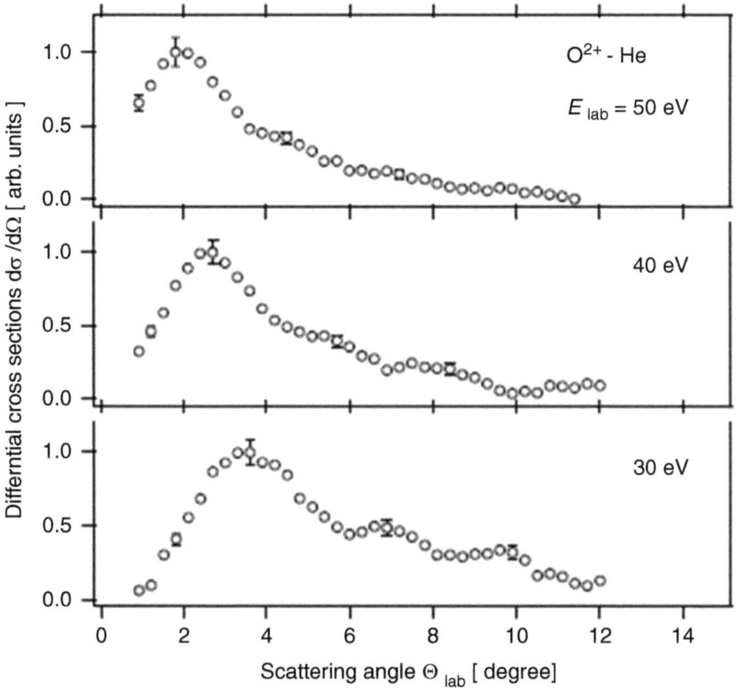

Fig. 3.26 Differential cross sections for single electron capture in O^{2+} + He collision, as a function of the scattering angle, for three projectile energies [10]

There is also a probability that the electron first stays on the target, then, after reaching R_o, the electron is captured by the projectile on the "way-out" (Fig. 3.25). These two ways lead to the same final state. Therefore, interferences may occur, leading to oscillations (called Stueckelberg oscillations) that can be observed.

The oscillations due to level crossing have been observed first in the case of symmetric systems, such as He^+ + He [9]. For asymmetric systems, the study of Stueckelberg oscillations began 30 years ago. To illustrate this phenomenon, we will consider two systems consisting of low charged (O^{2+}) and highly charged (C^{4+}) ions colliding on He, for which single- and doubly-differential cross sections have been measured, respectively.

The collision system O^{2+} + He was studied in 2002 [10], and the differential cross sections for single electron capture were measured as a function of the scattering angle θ_c (Fig. 3.26), for different projectile energies E_p. The cross section is found to increase as θ_c increases from 0 to 2–3°, and then to decrease at larger angles, with clearly visible oscillations. The amplitude of the oscillations increases as E_p decreases.

The presence of a maximum at angles smaller than 4° is due to the so-called rainbow effect. This effect can be explained classically. We consider the extreme case A^{Z+} + H, where A has a mass larger than that of H. At an impact parameter

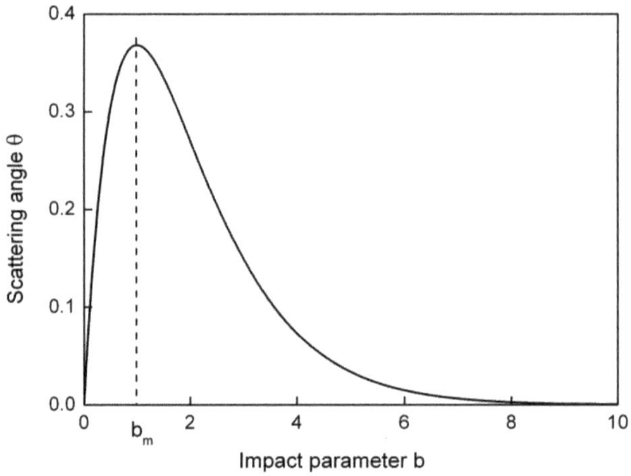

Fig. 3.27 Schematic dependence of the scattering angle on the impact parameter

Fig. 3.28 Dependence of the scattering cross section on the impact parameter. The value b_m corresponds to a scattering angle of 0.4°

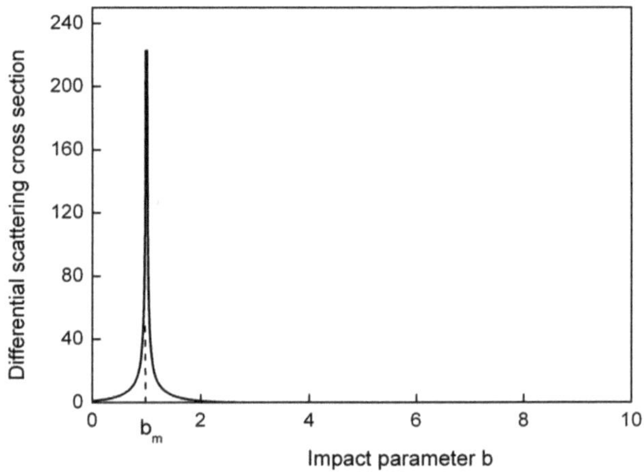

$b = 0$, the trajectory of A^{Z+} is a straight line, and the scattering angle is $\theta = 0$. At large impact parameters, the influence of H on A^{Z+} is negligible so that $\theta = 0$. Thus, there is a particular value b_m of b for which the deviation is maximal. Figure 3.27 shows schematic behavior for θ *(b)*, which presents a maximum at b =1 a.u.

Classically, the differential scattering cross section is $\frac{d\sigma}{d\Omega} = \frac{b}{\sin\theta}\left|\frac{db}{d\theta}\right|$ (Appendix 2). Thus, the cross section has a maximum value at $b = b_m$ (Fig. 3.28). This is the reason why the cross section is maximal for a scattering angle other than 0.

As shown theoretically [11], the oscillations at angles larger than 4° are due to the fact that at least two paths lead to the same final state (Stueckelberg oscillations). The result of the calculation is shown in Fig. 3.29 (full curve) and

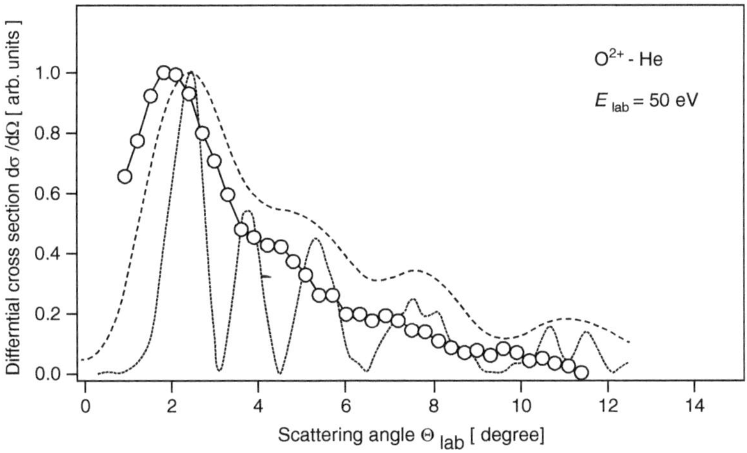

Fig. 3.29 Single capture differential cross section as a function of the scattering angle for the O^{2+} + He system at 50 eV. Experiment: *open circles*; calculation [11]: *full curve*; calculation including convolution with angular resolution: *dashed curve*

compared with the experiment at a projectile energy of 50 eV. To make a more efficient comparison with the experiment, the calculated cross section was then convoluted with the experimental angular resolution (dashed curve). The agreement produced is quite good, since the position of the maxima between 4° and 9° are similar. We note that the position of the first maximum due to the rainbow effect is shifted by ~0.5°, compared with the experimental one.

In the case of C^{4+} + He collisions, doubly differential capture cross sections were measured at energies ranging from 240 eV up to 440 eV. The cross sections were evaluated as a function of the energy gain ΔE, which is the difference in projectile kinetic energy before and after the collision, and the scattering angle. Some typical 3D-spectra are presented in Fig. 3.30. The solid and broken curves show the kinematic relations between ΔE and θ for the two different states $C^{2+}(1s^2 2s^2\ {}^1S)$ and $C^{2+}(1s^2 2s2p\ {}^1P)$, respectively. The formation of $C^{2+}(1s^2 2s2p\ {}^1P)$ is found to be dominant. The 3D representation enables us to clearly resolve the oscillation patterns that correspond to the Stueckelberg oscillations. The integration of the cross section over ΔE considerably reduces the amplitude of the oscillations.

3.4 Electron Emission in Atomic Collisions: Young 1801

3.4.1 Photoionization of Molecules

Since the pioneering work by Cohen and Fano in 1966 [12], many experimental and theoretical studies have searched for interferences in ionization of molecules by photons. In their paper [12], Cohen and Fano focused their attention on an

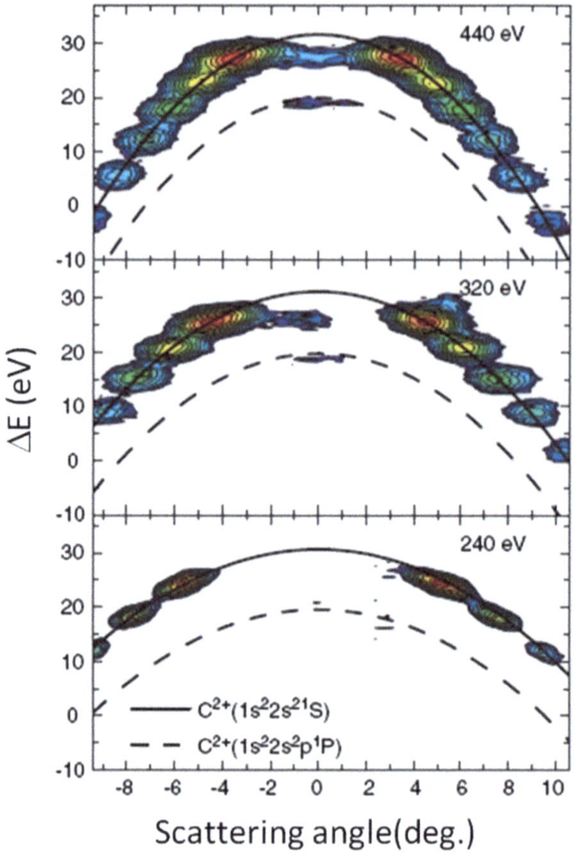

Fig. 3.30 Doubly-differential cross sections of scattered C^{2+} ions in C^{4+} + He collisions at 240, 320 and 440 eV. *Solid and broken curves* indicate the expected ΔE as a function of the scattering angle for the final states $1s^22s^2\ ^1S$ and $1s^22s2p\ ^1P$

experiment [13] where photoionization cross sections were measured for N_2 and O_2 molecules as a function of the incident energy of the photon. Figure 3.31 shows typical measured cross sections for these two molecules.

At photon energies below 30 eV, some structures are present, probably reflecting detailed properties of valence orbitals. At higher energies, the cross section decreases strongly with increasing photon energy. This trend is expected according to both qualitative theory and experiments on comparable single atoms.

At photon energies larger than 35 eV, structures appear, described as *shoulders* by the authors. They interpreted these undulations from Huygens' point of view. The two atoms of N_2 and O_2 molecules constitute two separate sources of electrons (Fig. 3.32). Superposition of the emissions from these two sources produces an interference pattern whose properties depend periodically on the photon wavelength and the internuclear distance.

Suppose that the emitted electrons are assigned to various sets of continuum states, classified by quantum numbers, such as the angular momentum ℓ. Cohen and Fano suggested that photoabsorption by transitions to states with increasing ℓ

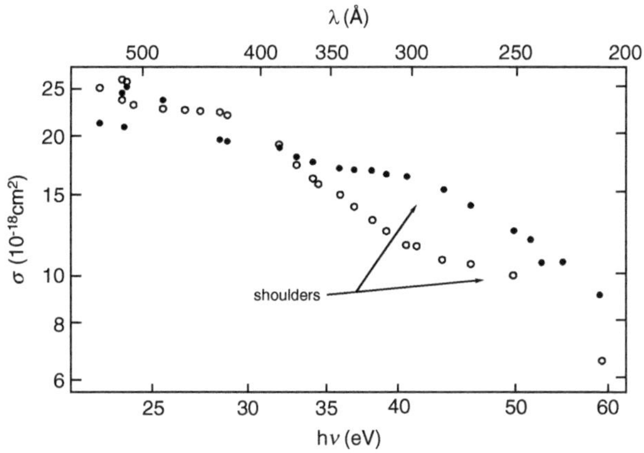

Fig. 3.31 Photoionization cross sections of the N_2 and O_2 molecules, as a function of the photon energy [13]

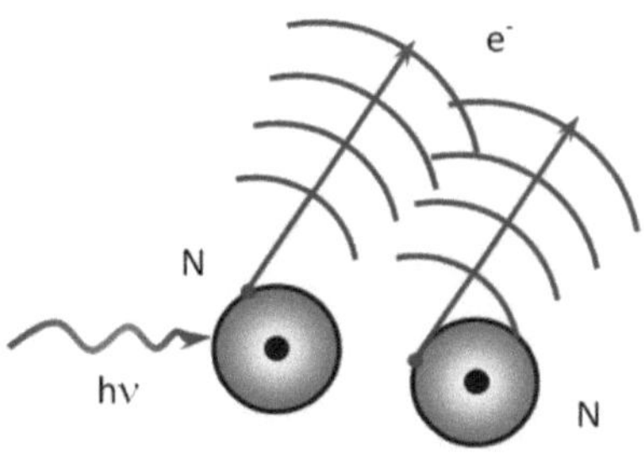

Fig. 3.32 A photon of well-defined energy hν ionizes a molecule, which consists of two independent coherent sources. The overlap of the waves produces interferences

becomes appreciable at increasing photon energies. They estimated that the value of $\ell = 3$ is relevant to the photon energy range 30–60 eV. To gain further insights, the cross section for photoabsorption was calculated for H_2^+ with the Born approximation. Expanding the final state wave function in terms of free-field spherical waves, the total cross section can be expressed as

$$\sigma = \sum_{\ell} \sigma_{\ell} \tag{3.9}$$

where σ_ℓ is the partial cross section. The result is presented in Fig. 3.33 as a function of the quantity $h\nu - I$, where $h\nu$ is the photon energy and I is the ionization energy of the hydrogen molecule for ℓ values ranging from 1 to 7. The total cross section is found to show undulations with a maximum at 200 eV, caused by the dominance of the $\ell = 3$ contribution.

The summation over all the ℓ values gives rise to the very simple formula

$$\sigma = \frac{1}{1.46}\,\sigma_H(Z^*)\left(1 + \frac{\sin kR}{kR}\right) \tag{3.10}$$

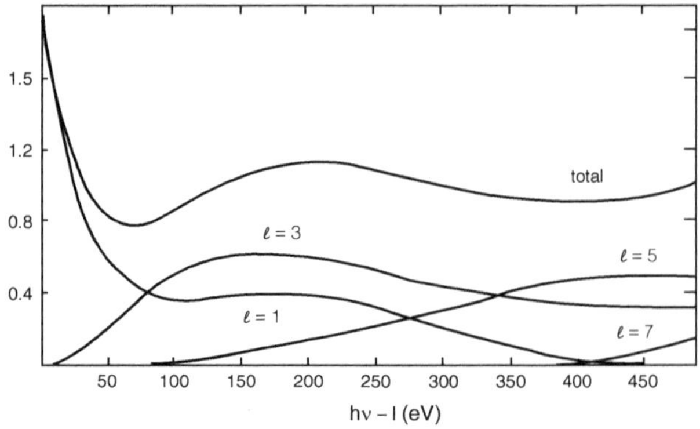

Fig. 3.33 Contribution of the different ℓ values to photoionization, as a function of the difference hν − I, and total contribution, revealing a maximum due to $\ell = 3$

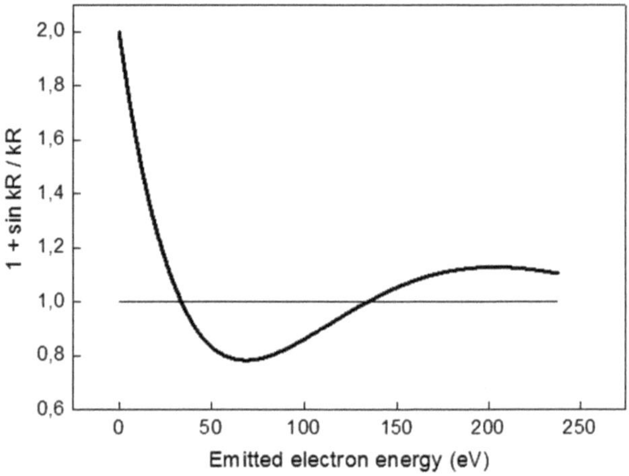

Fig. 3.34 Interference term $1 + \sin kR/kR$ plotted as a function of the emitted electron energy

In this formula, $\sigma_H(Z^*)$ is the photoabsorption cross section for a hydrogen-like atom of adjusted atomic number Z^*, k is the momentum of the emitted electron and R is the internuclear distance between the protons. The authors pointed out that the term $1 + \sin kR/kR$, plotted in Fig. 3.34 as a function of k, appears in different fields of physics, such as the acoustic power radiated by two point sources, or the scattering of X rays and massive particles by diatomic particles.

It can be seen that the observation of such an oscillation is not straight forward, since the period of the oscillation is large, and because the cross section at large photon energies is expected to be small.

Two years later, after the publication of Cohen and Fano's work, Bates and Öpik performed calculations for H_2^+ and for more complex molecules [14]. Surprisingly, in the case of H_2^+, no shoulder was observed, contrasting with Cohen and Fano's results. However, for complex molecules such as N_2^+, $\Delta\ell = 3$ and 5 contributions are too weak to explain the shoulder (Fig. 3.35). In fact, the shoulder is produced by the contribution from $\Delta\ell = 3$ and $\Delta m = \pm 1$ transitions (m is the projection of ℓ on a quantization axis), after a passage through a minimum value.

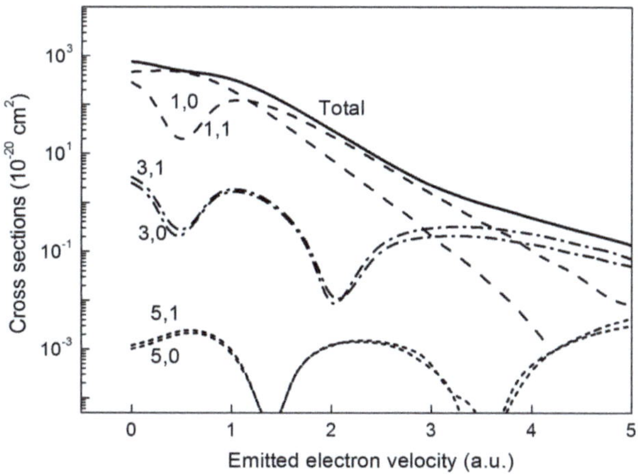

Fig. 3.35 Photoionization cross section, as a function of the emitted electron velocity, for different couples of values (ℓ,m). The total cross section (*full curve*) is the sum of the partial cross sections. The shoulder is found at about 1 a.u

With the appearance of coinciding measurements, progress was made in the understanding of the so-called *electron localization and de-localization*. In 2005, a molecular double-slit experiment using N_2 molecules was published [15]. In this experiment, molecules with undistinguishable ^{14}N–^{14}N atoms and partially distinguishable ^{14}N–^{15}N atoms were used. Thus, when using N_2 with equivalent atoms, the photoionization experiment resembles Young's double-slit experiment, while the introduction of ^{15}N is equivalent to partially marking one of the slits in Young's experiment. This provides partial information on which path is taken by an electron.

The experimental set-up is shown in Fig. 3.36. The soft X rays that ionize core electrons were provided by synchrotron radiation. All the photoelectron and fragment ion momenta were determined in coincidence, yielding the photoelectron angular dependence of fixed-in-space molecules.

Because both N atoms have two 1s electrons, the N_2 molecule has 1σ orbitals that are split into two orbitals, called $1\sigma_g$ and $1\sigma_u$ (g: gerade or even; u: ungerade or odd), in order to respect Pauli's principle. These orbitals are separated by an energy of 97 meV. The small ground-state splitting indicates the existence of a tunneling barrier between the two cores of the molecule, which allows a partial delocalization of the electrons between the two atoms.

With the spectrometer described above, the authors were finally able to settle the issue of electron non-locality. Figure 3.37 shows the angular dependence of emitted electrons in a two dimension plot for the cases of $1\sigma_g$ (top) and $1\sigma_u$ (middle). The difference between these dependences (bottom) clearly shows an interference pattern, revealing the coherence of the emission.

It should be noted, however, that, contrary to what the authors say, their experiment does not quite resemble that of Young published in 1807. In Young's experiment, the light source was independent of the interferometer while, in all the experiments described above, the electron is emitted from the interferometer itself. In fact, the experiments are more like the assertion by Young that the experiment. He postulated that the emission of waves by two independent sources give rise to maxima and minima in the amplitudes.

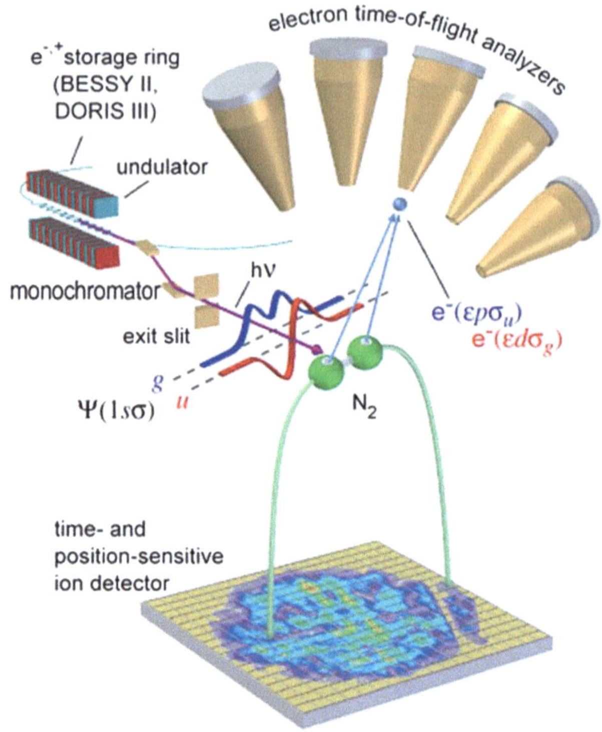

Fig. 3.36 Experimental set-up for photoionization of N_2 molecules. The fragment ions and the emitted electron are detected simultaneously. The angular distribution of electrons is determined

Another way to reveal interferences is the comparison of cross sections after photon impact with a molecule, for example H_2, and the associated atom, i.e., H in this example. It should be remembered that the photoionization cross sections strongly decrease as the photon energy increases (Fig. 3.38).

The cross sections may vary by several orders of magnitude according to energy, whereas the variation due to interference effects is limited to a factor of two. Since photon impact on H is not expected to give rise to interferences, the ratio $\sigma(H_2)/2\sigma(H)$ would clearly reveal the wave character of the electron. As an example, the ratio $\sigma(H_2)/2\sigma(H)$ is shown in Fig. 3.39, for photon energies from the threshold up to 20 a.u. Because no experimental cross sections for H were available, the experimental ratio is in fact the ratio between the H_2 experiment [17] and H calculations [18]. This experimental ratio is found to increase at low photon energies, reaching a maximum of 1.8 at ~1 a.u. Then the ratio decreases to reach a minimum value of 1.3 at ~2.2 a.u.

In the figure, the theoretical results, based on an approach that takes into account the electron correlation and the molecular character (red curve) of H_2 molecular states reproduce the experimental ratio well. The simple formula by Cohen and Fano [12], using some free parameters, also agrees with the experiment, except at photon energies smaller than 1 a.u. In fact, as shown by the blue curve in Fig. 3.39, the decrease of the cross sections at low energies originates mainly from

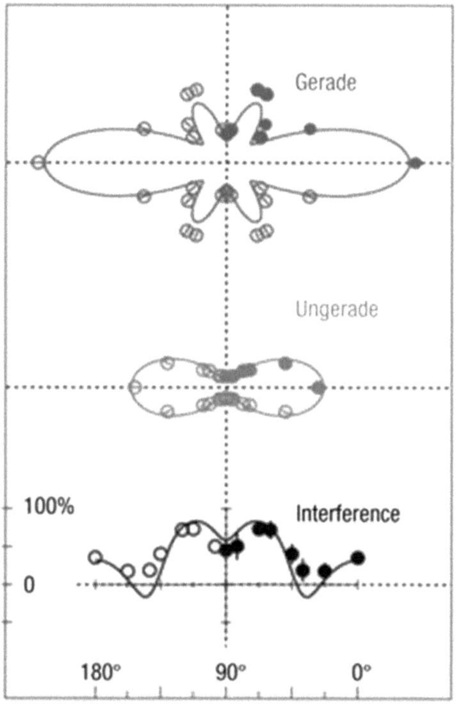

Fig. 3.37 Angular distribution of electrons emitted after photon impact when the molecule in its gerade (*top*) and ungerade (*middle*) state. The difference of the two distributions (*bottom*) shows oscillations due to interferences

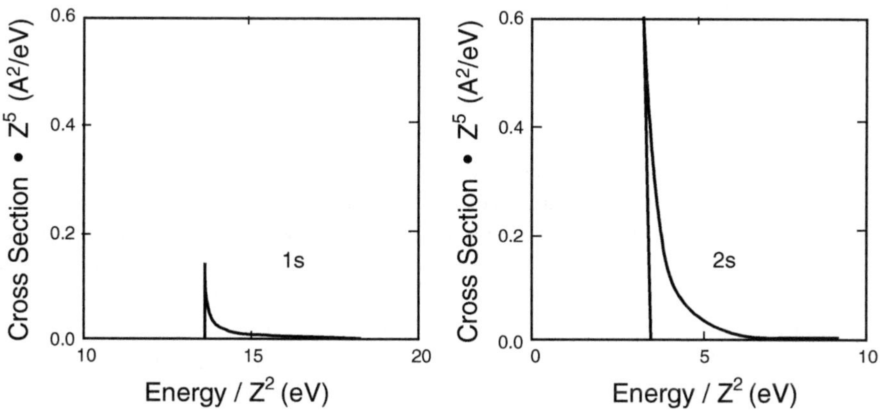

Fig. 3.38 Reduced photoionization cross sections for hydrogenic atoms as a function of the reduced photon energy (From [16])

correlation. Without including the electron correlation, and by considering the molecule as two distinct atoms, the ratio (blue curve) resembles the ratio obtained by Cohen and Fano.

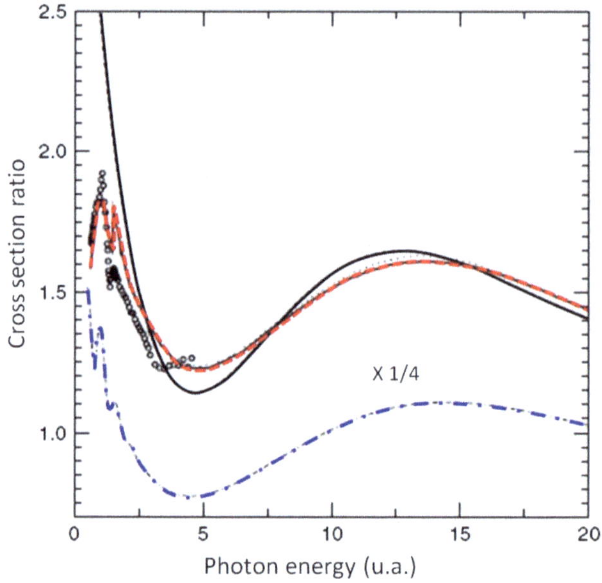

Fig. 3.39 Experimental (*open circles*) [17] and theoretical ratios $\sigma(H_2)/2\sigma(H)$ (*curves*). Black full curve: simple formula by Cohen and Fano; *red full curve*: ab-initio method results; *blue dashed-dotted curve*: results for a fictitious one-electron molecule with nuclear charges equal to 0.5 (for clarity, the result has been divided by 4)

3.4.2 *Ionization of H_2 by Ion Impact*

3.4.2.1 First Order Interferences

What happens if the incident photon is replaced by a charged particle, such as an ion or an electron? One can reasonably expect a similar effect, since the role of the incident particle is to extract an electron from the molecule. Nevertheless, some important differences need to be pointed out. In the case of a photon, the energy of the emitted electron is governed by the photon energy, since it is the difference between the photon energy and the energy binding the electron to the molecule. On the contrary, in the case of charged particle impact, the energy of the emitted electron strongly depends on the impact parameter. Therefore, a fixed incident particle energy E_p corresponds to a broad energy distribution of the emitted electron.

Interference effects following charged particle impact on H_2 were studied in the early 2000s. The first experiment was realized at Ganil (*Grand Accélérateur National d'Ions Lourds*) in Caen, France, using Kr^{34+} at 60 MeV/amu as projectiles [19]. The electrons were detected as a function of their energy in the range from 2 eV up to 200 eV, and for detection angles θ_d between 20° and 160° with respect to the incident beam direction.

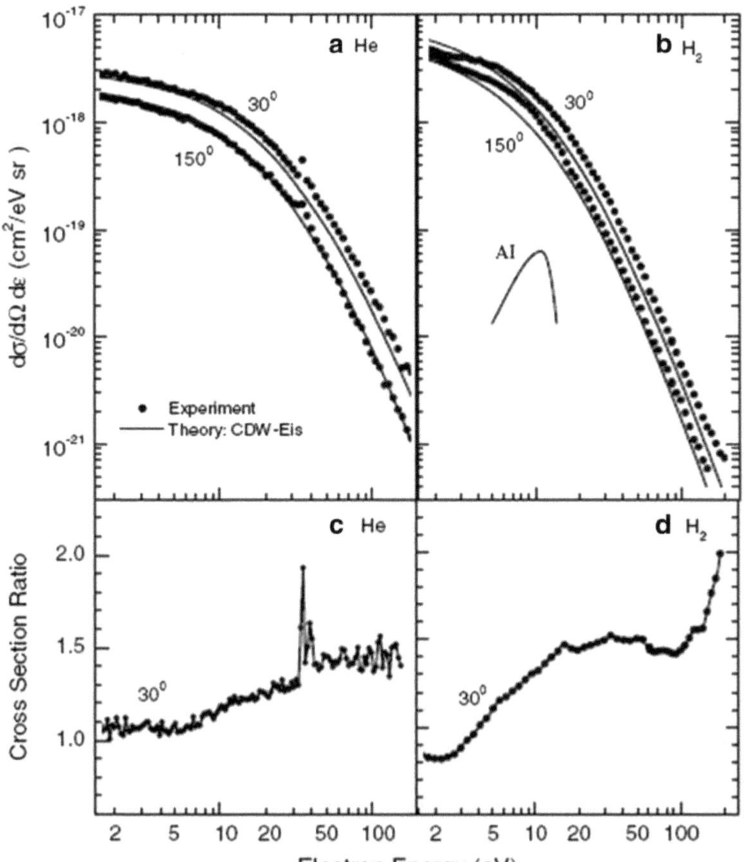

Fig. 3.40 *Top*: spectra realized at 30° and 150° for He (*left-hand side*) and H_2 (*right-hand side*). *Full curves*: result of theoretical calculations. *Bottom*: experimental to theoretical ratios

Two typical spectra are shown in Fig. 3.40, at symmetric angles of 30° and 150° (right side of the figure). For comparison, the experiment was also realized with He target (left side of the figure). Superimposed on the spectra, the results of calculations are also plotted (full curves). For H_2, calculations were performed by means of continuum-distorted-wave (CDW) approximation assuming independent H atoms, using hydrogenic wave functions with an effective charge $Z^* = 1.05$. In the case of He, the calculations were made using initial and final state Hartree-Slater wave functions.

As mentioned previously, since cross sections strongly decrease with increasing electron energy, the most useful way to enhance the visibility of possible interferences is to divide experimental cross sections by theoretical ones. The result of the division is given at the bottom of Fig. 3.40 for both targets. In the case of He (Fig. 3.40c), apart the structure at ~30 eV due to autoionization, the ratio increases

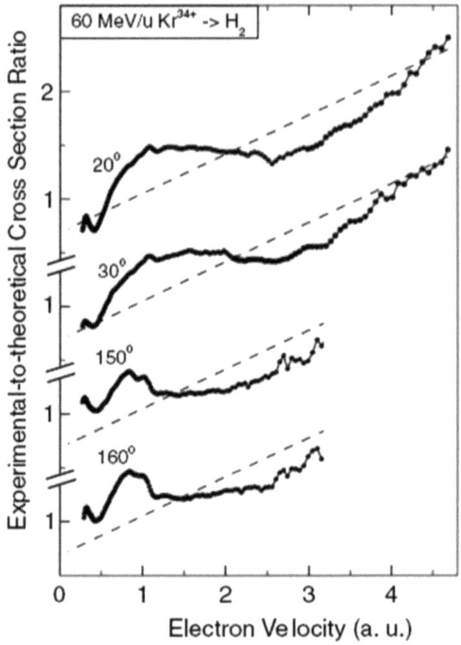

Fig. 3.41 Cross-section ratios at four detection angles. The *dashed lines* fit the overall increase

monotonically, reflecting the disagreement between experiment and theory at large electron energies. On the contrary, the ratio for H_2 increases non-monotonically, suggesting the presence of an oscillatory structure.

To allow for a direct comparison with theory, the overall increase of the ratio was fitted with straight lines (Fig. 3.41).

One can first see that the oscillations are present at all the mentioned angles. Nevertheless, it is clear that the period of the oscillations is higher at forward angles than at backward angles. The straight lines were used to remove the increase of the ratio. In addition, since the interference term depends on the electron momentum k, the ratio for the different angles is determined as a function of k. The result of the subtraction is given in Fig. 3.42, for angles of 30°, 60°, 90° and 150°. The results are also fitted using Bessel functions and compared with Born calculations.

The Born calculations reproduce the experimental ratio fairly well, except at momenta smaller than 1 a.u. At 90°, Born calculation again reveals a smooth oscillation that is not observed experimentally. At 150°, one observes a shift in the oscillation, since the theory predicts symmetry between forward and backward angles.

The Bessel function originates from the same considerations formulated by Cohen and Fano [12]. After averaging over all the orientations of the molecule, the doubly differential cross section (DDCS) in energy and angle can be expressed as

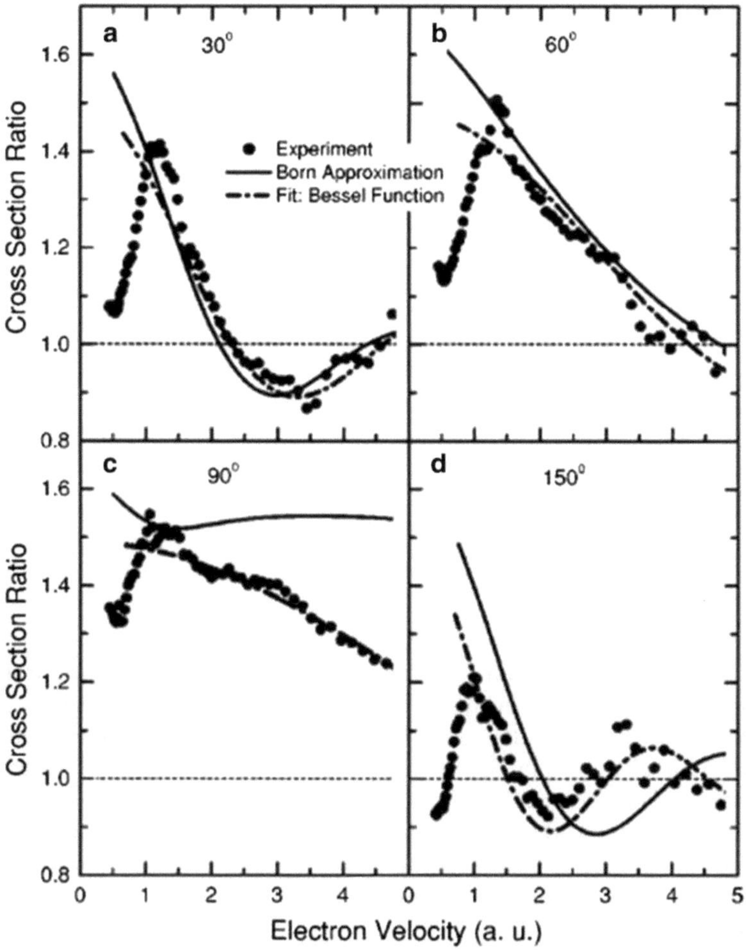

Fig. 3.42 Experimental to theoretical ratios (*black circles*) after subtraction of the *fitted curves* (see text), at angles of 30°, 60°, 90° and 150°. *Dashed curves*: Fit using Bessel functions. *Full curve*: Born approximation

$$\frac{d^2\sigma(H_2)}{dEd\Omega} \approx A\left(1 + \frac{\sin kd}{kd}\right) + B \tag{3.11}$$

where A and B are adjustable parameters, and d is the internuclear distance between the two protons at equilibrium ($d = 1.42$ a.u.); see Appendix 3 for further details. The result agrees very well with the experiment except, once again, at the lowest momenta. In the range between 0.2 and 1 a.u., it seems that the ratio exhibits oscillations that cannot yet be explained.

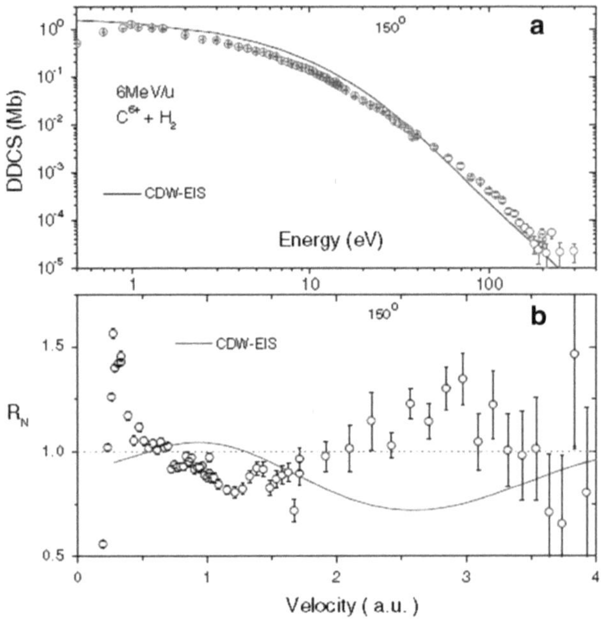

Fig. 3.43 DDCS measured for the system $C^{6+} + H_2$ at 150° [20] compared with the CDW-EIS calculations (*solid line*) and (**b**) the normalized DDCS ratio (R_N) showing oscillation around 1, (*dashed line*). The solid lines represent the fully calculated DDCS ratios

In 2004, direct evidence for interference effects has been discussed from electron emission spectra following ionization of H_2 with C^{6+} and F^{9+} ions [20]. First, as found previously for Kr^{34+}, the experiment was compared with CDW calculations. An example is given in Fig. 3.43 for 150°. At the top of the figure, experimental (open circles) and theoretical (full curve) DDCS are shown.

The experimental-to-theoretical ratio is shown on the bottom of the figure, and compared to theoretical-to-theoretical ratio. In both cases the oscillations are visible, but it is clear that the agreement between them is poor, since the ratios oscillate in counterphase.

As pointed out by the authors, the continuous increase in the ratio is due to the use of an incorrect value of the effective charge of the atomic hydrogen. Therefore, the shape of the oscillation is subject to the choice of theoretical parameters. To partly avoid this problem, the authors performed the experiment for both H_2 and H targets [20]. Figure 3.44 presents the experimental-to-experimental ratio $d^2\sigma(H_2)/2d^2\sigma(H)$ obtained with F^{9+} projectiles, as a function of the electron velocity, for detection angles of 45° and 60°.

At both angles, the ratio increases, reaches a maximum at ~4 a.u., then decreases. The theoretical shape (full curves) derived from CDW calculations is now in better agreement with the experimental one. However, as seen for 45°, the maximum of the theoretical ratio is obtained at an electron velocity of ~2 a.u. For comparison, the experimental-to-theoretical ratio is also shown (dashed-dotted curve on bottom of Fig. 3.44). This ratio strongly deviates from the experimental one, indicating that the present method is necessary, not only to reveal the oscillations, but also to determine the amplitude and frequency with higher precision.

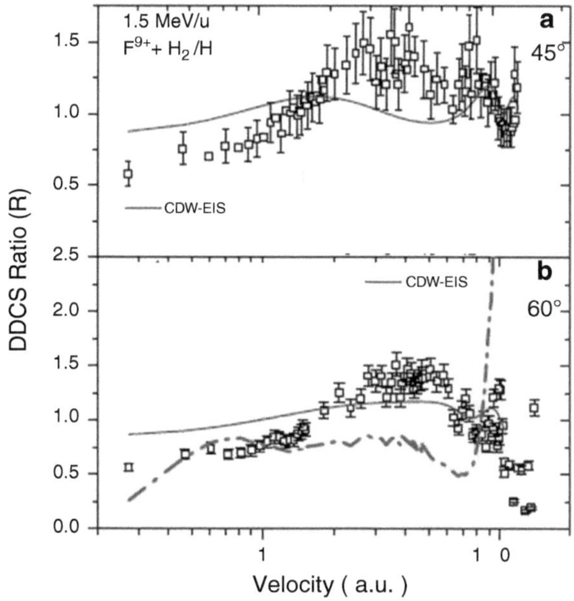

Fig. 3.44 Fully measured DDCS ratio for F^{9+}at detection angles of (**a**) 45° and (**b**) 60°, compared with the fully calculated DDCS ratios (*solid lines*). The experimental-to-theoretical ratios (*dash-dotted line*) are also shown for comparison

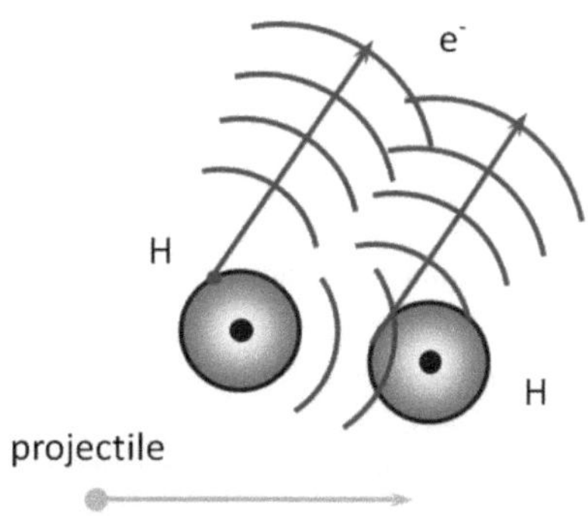

Fig. 3.45 The wave associated with the emitted electron scatters on the proton (from the left to the right) before being emitted. Before being emitted and detected, an infinity number of movements of back and forth are possible between the protons

3.4.2.2 Multiple-Order Interferences

To explain the interference pattern, we assumed that the waves originating from the protons overlap (Fig. 3.32). In fact, one has to take into account the possibility for a wave originating from one proton to partly scatter on the other proton before the emission (Fig. 3.45). Thus, before being emitted, the wave associated with the electron may scatter several times, with probabilities that decrease as the number of scatterings increases.

The effect of multiple-order interference phenomena in ion-atom and ion-molecule collisions has been the subject of many studies and controversial discussions. In 2004, the system $Kr^{34+} + H_2$ was revisited to search for evidence of

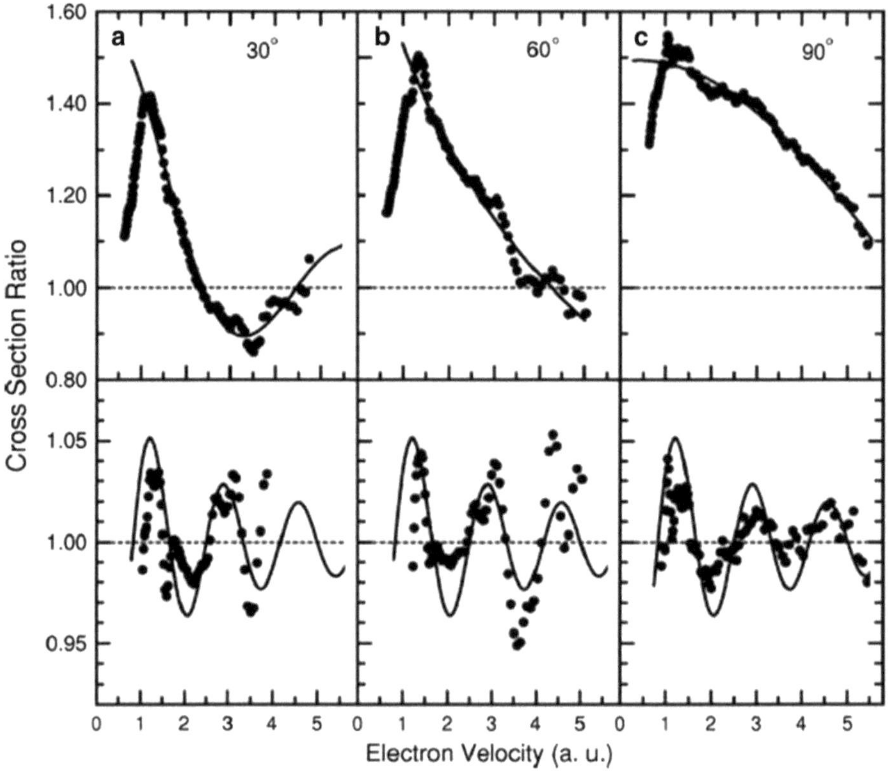

Fig. 3.46 Experimental to theoretical ratios, at three angles. The ratios are divided by their corresponding fits (*full curve*)

multiple-order effects [21] by analyzing the cross-section ratio in greater depth (Fig. 3.46).

The experimental-to-theoretical ratio R_N, shown in Fig. 3.46, is divided by the function that fits the ratio (full curve). The result of the division, presented on the bottom of the figure, shows oscillations whose frequency is twice the main frequency. This frequency, whose amplitude is of the order of 0.05, is seen to be independent of the detection angle.

The present effect, attributed to second order effects, has been also revealed in $F^{9+} + H_2$ [22] and $H^+ + H_2$ [23] collisions. In $F^{9+} + H_2$ collisions, the experimental-to-theoretical ratio (Fig. 3.47) was fitted using the function

$$R = A(\theta) + B(\theta)\left(1 + \frac{\sin c(\theta)kd}{c(\theta)kd}\right) + D(\theta)\left(1 + 2\cos 2kd\,\frac{\sin kd}{kd}\right) \qquad (3.12)$$

In this relation, the first two terms describe the main oscillation, with adjustable parameters $A(\theta)$ and $B(\theta)$, while the last term is supposed to include the second-

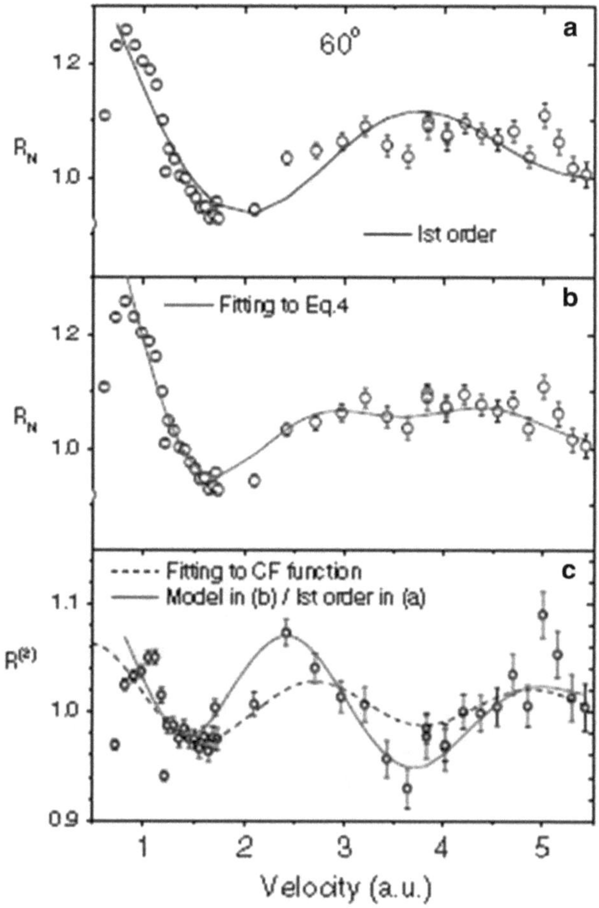

Fig. 3.47 Experimental-to-theoretical ratio for the collision system $F^{9+} + H_2$, at a detection angle of 60°. (**a**): the ratio is fitted using a Bessel function. (**b**): the third term of relation 3.12 is included. (**c**): the ratio is divided by the Bessel function that fits it (*full curve in a*). The *dashed curve* is obtained from the Cohen-Fano model

order interferences [21]. The quantity $c(\theta)$ is a fitting parameter whose form depends on the detection angle, as will be discussed later.

If the third term of relation (3.12) is not used in the fitting procedure (Fig. 3.47a), the ratio R_N is not well reproduced. The third term is found to be necessary (Fig. 3.47b) to correctly fit the ratio. However, even if the third term is included, the result of the fit does not explain the increase of the ratio at velocities increasing from 0 to 1 a.u.

A complementary study was performed in 2005 using the collision system $H^+ + H_2$, at projectile energies of 1, 3, and 5 MeV [23]. Again, the experimental-to-theoretical ratio was determined (Fig. 3.48) for several detection angles (three examples are given on the figure), and was fitted using a function of the form $A(\theta) + B(\theta)\ (1 + \sin c(\theta)kd/c(\theta)kd)$ (full curve in the figure), where A, B and c are adjustable parameters. In addition, the ratios were compared with Born calculation results. As in the case of the other collision systems, the calculations agree with the experiment, except at backward angles, since the calculated frequency is smaller than the experimental one.

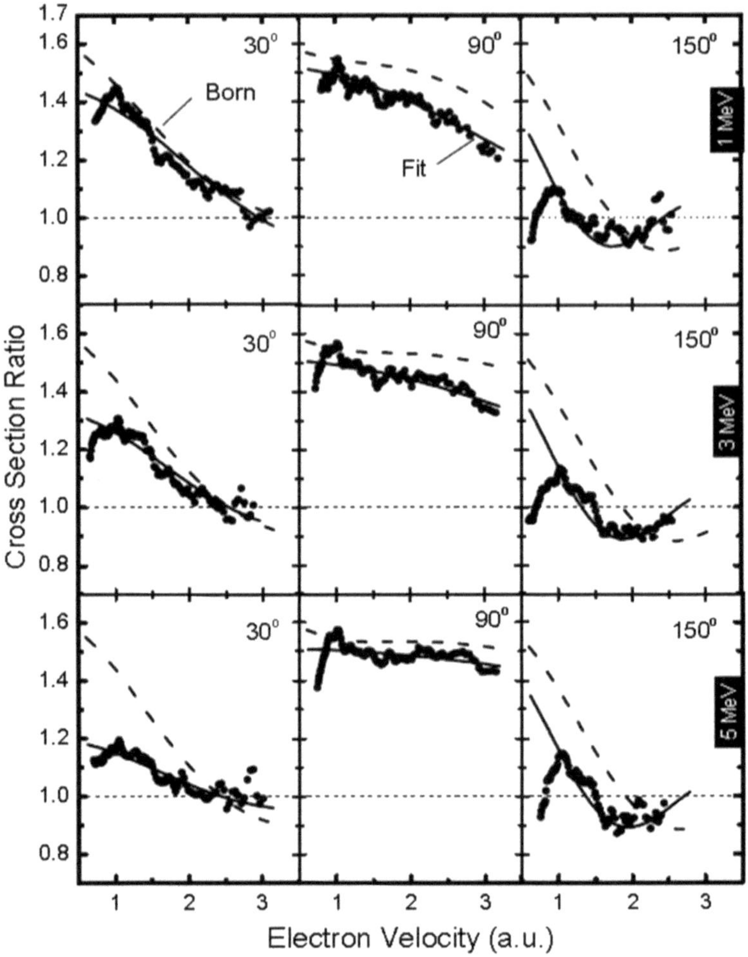

Fig. 3.48 Experimental ratios (*black circles*) at energies of 1–5 MeV, and for detection angles of 30°, 90° and 150°. The *full curve* fits the ratio for electron velocities larger than 1 a.u., while the *dashed curves* show the result of Born calculations

A close inspection of the ratios suggests the existence of higher-frequency structures superimposed on the main oscillatory structure. To display this structure more clearly, the ratios have been divided by their corresponding fits to give the results shown in Fig. 3.49. As in the case of Kr^{34+} + H_2, the resulting ratios exhibit secondary oscillations with double frequencies dashed sinusoidal curves). More surprisingly, Fig. 3.49 suggests the existence of additional structures with much higher frequencies. It should be noted that, in contrast to the first-order oscillation, neither the secondary nor the high-frequency structures appear to be dependent on the collision velocity (Fig. 3.49a) or the detection angle (Fig. 3.49b).

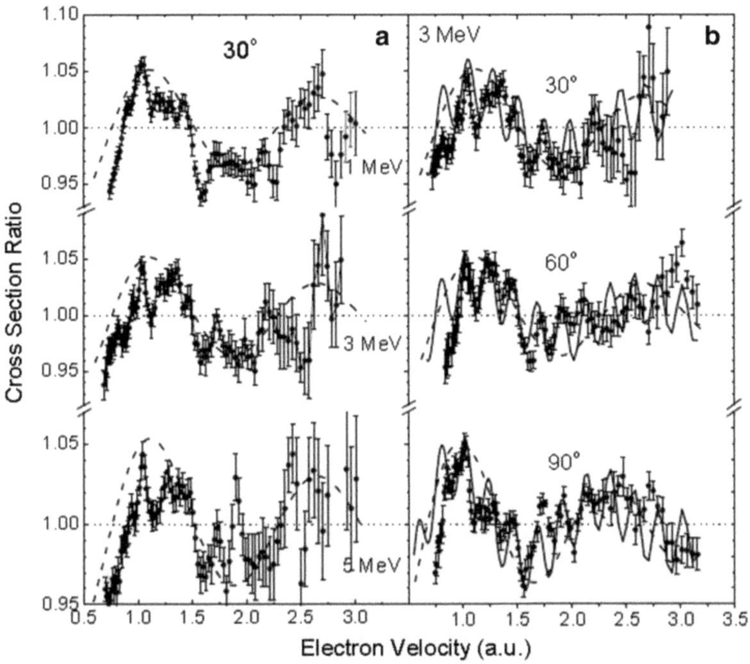

Fig. 3.49 Experimental-to-theoretical ratio of Fig. 3.48 with their corresponding *fitted curves*. *Left-hand side*: ratio at 30° and three projectile energies. *Right-hand side*: ratio at 3 MeV and three detection angles

The origin of these high-frequency structures is not immediately clear. Due to the high velocity of the projectile, coherent emission from a transient molecule formed by a target proton and the projectile proton is unlikely. These oscillations may be caused by the interference between direct electron emission (ionization) and the electron emission due to target autoionization following double excitation of H_2 into doubly excited states, mainly $2\ell n\ell'$ ($n \geq 2$) configurations.

3.4.2.3 Theoretical Calculations Including Multiple-Order Effects

After the discovery of interference effects in the ionization process following ion impact, several models were proposed to describe the observed oscillations in the electron velocity distribution. The triple differential cross section (TDCS) for electron emission is expressed as

$$\frac{d^3\sigma(H_2)}{dqdEd\Omega} = \frac{d^3\sigma(2H)}{dqdEd\Omega}\left(1 + \cos\vec{p}.\vec{d}\right) \tag{3.13}$$

In this expression, $\vec{q}$ is the momentum transfer, defined as the difference between the initial and final projectile momentum, the solid angle $d\Omega$ and energy $d\varepsilon$ refer to the ejected electron, and $\vec{p}$ is defined as $\vec{p} = \vec{k} - \vec{d}$, where $\vec{k}$ is the electron momentum, and $\vec{d}$ is the internuclear distance between the two protons. Since the molecular orientation is unknown, the cross section is averaged over the orientation of the internuclear axis, characterized by $\vec{d}$.

$$\frac{d^3\sigma(H_2)}{dqdEd\Omega} = \frac{d^3\sigma(2H)}{dqdEd\Omega}\left(1 + \frac{\sin pd}{pd}\right) \tag{3.14}$$

In the 2001 experiment, the quantity q was not measured, so that the TDCS had to be integrated over q

$$\frac{d^2\sigma(H_2)}{dqdEd\Omega} = \int \frac{d^2\sigma(2H)}{dqdEd\Omega}\left(1 + \frac{\sin pd}{pd}\right) dq \tag{3.15}$$

Following Bethe [24], the atomic cross section was written as a sum of two independent terms. One term corresponds to the dipole transition (soft collisions) whereas the second contains the contribution of the binary encounter process (hard collisions). Consequently, the cross section can be expressed as $A + B \sin kd/kd$. However, this relation is not angle dependent, while the experiment shows a strong dependence of the detection on the cross-section ratio. A new theory, carried out in the framework of semi-classical approximation, was performed in 2002 [25]. The authors obtained the following simplified cross section that takes into account the angular dependence

$$\frac{d^2\sigma(H_2)}{dEd\Omega} = A + B\frac{\sin\left(k_{//} - q_{\min}\right)d}{\left(k_{//} - q_{\min}\right)d} \tag{3.16}$$

Here, $k_{//} = k\cos\ \theta$ is the component of the electron momentum parallel to the ion beam, and $q_{\min}$ is the minimum momentum transfer, for which a good approximation can be obtained by $q_{\min} = \Delta E/v_p$, where ΔE is the energy transfer and v_p is the projectile velocity. At high v_p, $q_{\min}$ tends towards 0, so that the frequency varies approximately with $\cos\theta$. This dependence is confirmed experimentally except at angles larger than 90°, since the experimental frequency is much larger than the calculated one.

Classical Trajectory Monte-Carlo (CTMC) methods [26], as well as CDW calculations [27], have also been used to interpret the experimental oscillations. In all cases, the agreement is rather poor. For example, using CDW approximations for the collision involving Kr projectiles, (Fig. 3.50), the agreement is correct at 30° and, to a lesser extent, at 60°. At 90°, the theory totally disagrees with the experiment, while it qualitatively reproduces it at 150°. At this angle, a shift in the oscillation is observed.

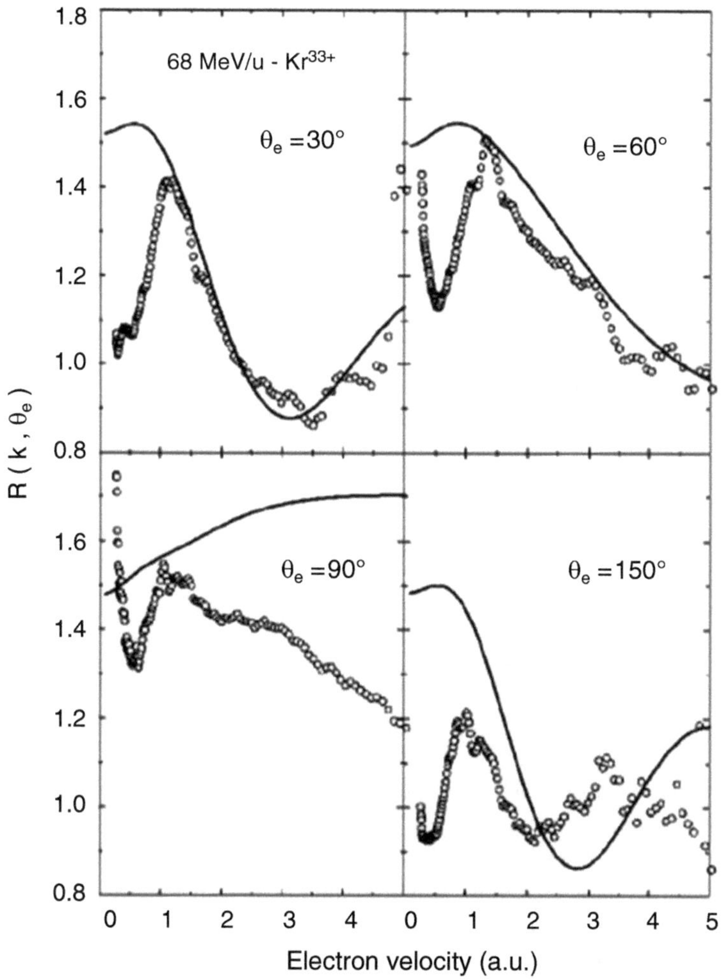

Fig. 3.50 Experimental (*open circles*) [19] and theoretical (*full curves*) [27] ratios of H_2 and H cross sections for Kr^{34+} projectiles, at detection angles of 30°, 60°, 90° and 150°

According to the authors, theoretical CDW calculations are not expected to give a good description of experimental cross sections at electron velocities lower than 1 a.u., since it is necessary to use accurate bound and continuum electron target wave functions. In fact, as shown in the case of photoionization [18], the abrupt fall of the experimental level, as velocity decreases, is not due to interferences, but to electron correlation and/or screening effects.

In 2004, an attempt was made to take into account second-order interference effects [28]. The calculations, based on the impact parameter formulation, were performed for interferences observed in $H^+ + H_2$ collisions. The double scattering

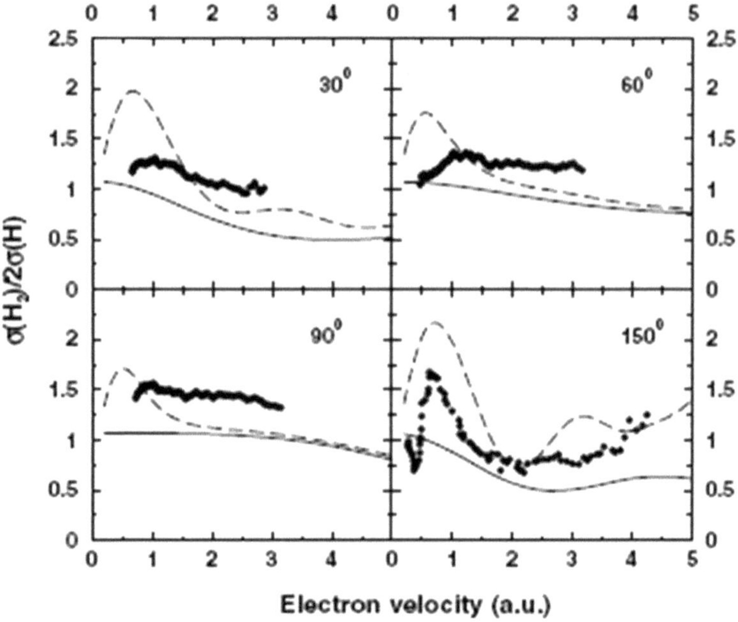

Fig. 3.51 Experimental-to-theoretical ratio (*full circles*) and theoretical ratios (*curves*) for H^+ + H_2 collisions. *Dashed curve*: result of first-order calculations. *Full curve*: Result of the sum of first- and second-order calculations

of the emitted electron was investigated. First, the electron is supposed to make a transition from the initial state to an intermediate state caused by the interaction with the projectile. Then, the electron is scattered by the protons. The intermediate and final states, characterized by $\overrightarrow{k_i}$ and $\overrightarrow{k_f}$ momenta, respectively, are described by plane waves, while the initial state is taken as a linear combination of two hydrogen-like 1s atomic orbitals. The ratio $\sigma(H_2)/2\sigma(H)$ is finally determined as a function of the electron velocity. A typical example is shown in Fig. 3.51, for a projectile velocity of 3 MeV and detection angles of 30°, 60°, 90° and 150°. In addition to the experimental-to-theoretical ratio (full circles), first-order (full curves) and the sum of first- and second-order results (dashed curves) are plotted.

The main contributions of second-order calculation are an enhancement of the ratio at all the detection angles, as well as a decrease of the ratio when the velocity decreases below 1 a.u., qualitatively reproducing the experiment. At 30° and 150°, the frequency is seen to be doubled compared with the first-order result. On the contrary, no frequency doubling is observed at 60° or 90°. Finally, the existence of a clear signature of multiple-order interference effects in ion molecule collisions is not proved.

To reduce the difficulty of associating the observed additional structures in the ratio with multiple-order scattering effects, some simplifications can be made. Suppose, to simplify, that scattering of order ℓ gives rise to interference terms of

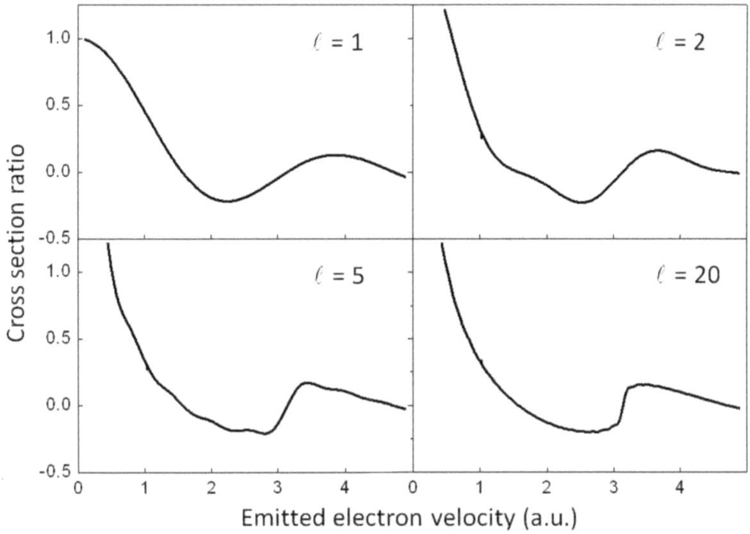

Fig. 3.52 Cross-section ratio calculated using the simple formula (3.17), for various values of interference order ℓ. For $\ell = 1$, one oscillation is observed. The second order reveals a double frequency oscillation. For large orders, only one oscillation remains

the form $\sin \ell kd/kd$, with a probability decreasing when increasing ℓ, such as $1/\ell$. Therefore, the total ratio can be written as

$$R = \sum_{\ell=0}^{\infty} \frac{\sin \ell kd}{\ell kd} \tag{3.17}$$

The result for R is plotted in Fig. 3.52, for several values of ℓ. The second order gives rise to frequency doubling, but the increase in the number of multiple scattering events tends to cancel the secondary oscillations (Fig. 3.52 for $\ell = 20$).

High frequency oscillations were suggested to originate from coherent emission from the transient molecule formed by the passing projectile ion and one of the H_2 target centers [23], as schematized in Fig. 3.53. In this case, such interference effects would also be expected in collisions of ions with atomic targets, where the number of atomic centers of the transient molecule is now reduced to two. In addition, such interference effects are likely to be more pronounced if the projectile and target centers are the same.

As an example, interferences between the target-centered and projectile-centered amplitudes were added to a two-center expansion model in order to describe the angular distribution of the ejected electrons in H^+ + H collisions at 20 keV [29].

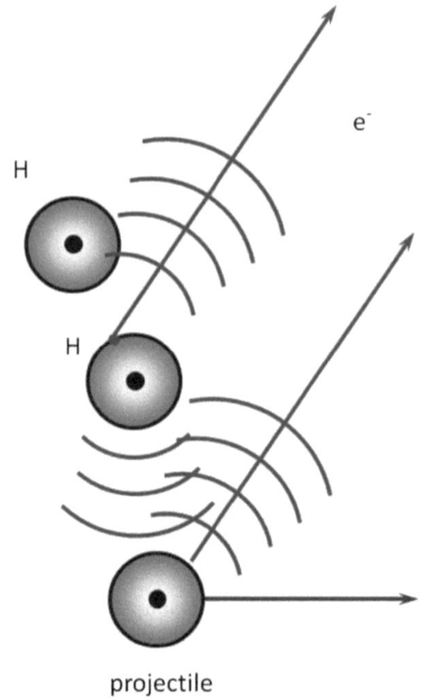

Fig. 3.53 Schematic interference pattern produced between the directly emitted target electron, and the electron that has first scattered on the projectile

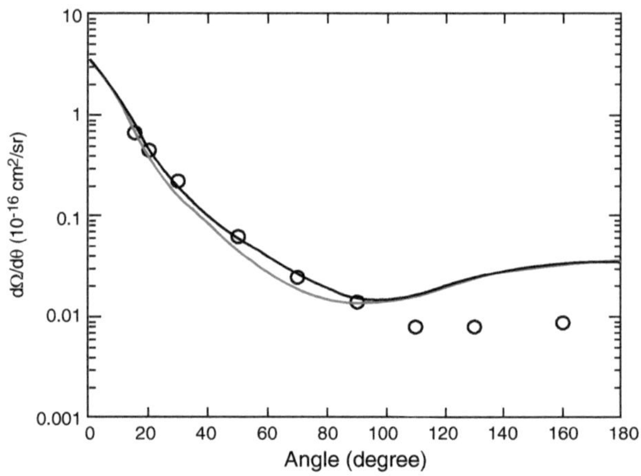

Fig. 3.54 Differential cross sections for ionization in 20 keV H^+ + H collisions. Experiment: *open circles* [30]. *Full curve*: theory (see text)

Experimental differential cross sections [30] are plotted as a function of the detection angle in Fig. 3.54 (open circles). The cross section strongly decreases from 0° to 110°, and remains rather constant at larger angles.

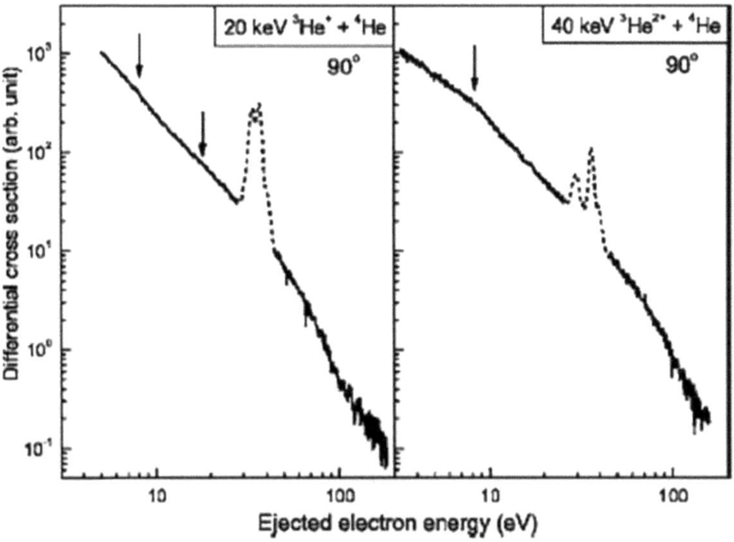

Fig. 3.55 Electron spectra obtained at 90° for He^+ + He (*left-hand side*) and He^{2+} + He (*right-hand side*) collisions. The structures are centered at ~30 eV

In Fig. 3.54, the results of the calculations are compared with the experiment. When the amplitudes are added incoherently (dotted curve), the agreement is good, except at angles larger than 90° where the calculated cross section is larger than the experimental one by a factor of ~3, as expected in such a model. The introduction of coherence between the projectile and the target improves the agreement with the experiment (full curve). Consequently, it seems that interferences between direct ionization and electron emission following scattering on the projectile play a non-negligible role in ionization.

More recently, ionization in He^{q+} + He collisions (q = 1, 2) has been investigated at projectile energies of 20 qkeV. Instead of studying the angular distribution of emitted electrons, the signature of possible high-frequency interferences has been searched for in electron energy distributions at a fixed detection angle. Figure 3.55 shows two typical DDCS (solid lines) at 90°, for He^+ + He (left) and He^{2+} + He (right) collisions.

The cross section is found to decrease by four orders of magnitude in the range 4–200 eV. The structures superimposed on the ionization background are caused by double excitation of the target and double capture in the case of He^{2+} + He collisions. The small broad structures, indicated by the arrows, are also seen, and are systematically observed at all the studied angles. As in the case of collisions between ions and H_2 molecules, the present experimental cross sections were divided by a function that reproduces the ionization contribution. Since no calculation is available for these systems, DDCS were divided by a polynomial function in log-log scale, whose degree varies from 2 to 4. To verify the validity of the procedure, the same technique was applied to the e^- + He system, for which no oscillation is expected.

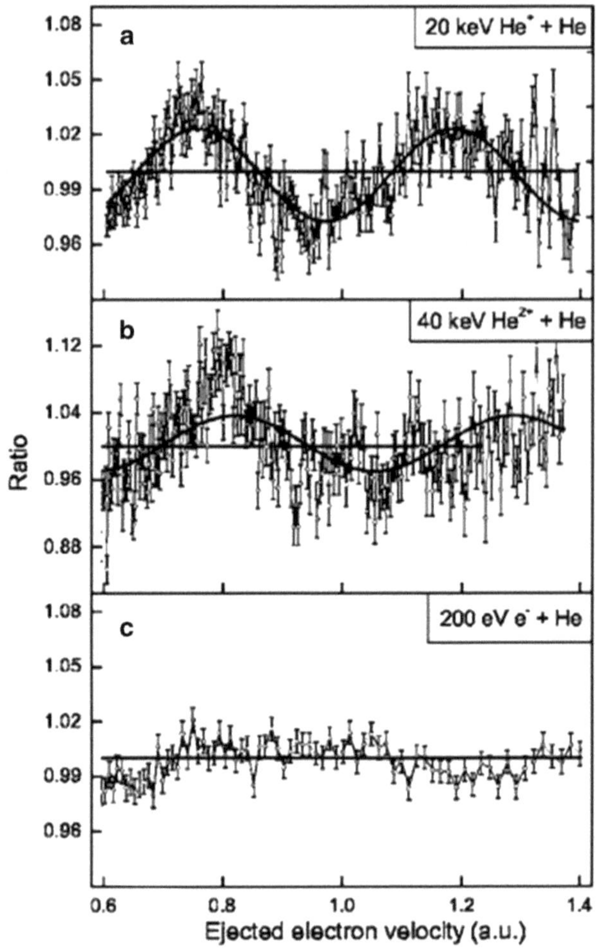

Fig. 3.56 Experimental ratios of DDCS and polynomial fits for He^+ + He (**a**), He^{2+} + He (**b**) and e^- + He (**c**), as a function of the emitted electron velocity

The results of the division are shown in Fig. 3.56, for He^+ + He (Fig. 3.56a), He^{2+} + He (Fig. 3.56b) and e^- + He (Fig. 3.56c). While no oscillation is visible in the latter case, two clear oscillations are shown in He^{q+} + He collisions, with an amplitude of ~0.03, well above the uncertainties, which are less than 1 %. To better visualize the oscillations, the ratios were fitted with sinusoidal functions (full curves). The period of the oscillations, of the order of 0.4 a.u, is found to be rather independent of the collision system and the detection angle. Surprisingly, this period is of the same order of magnitude as that found for H^+ + H_2 collisions [23]. The present result supports the hypothesis that interferences between direct ionization and emission after projectile scattering should be taken into account in the analysis of electron spectra.

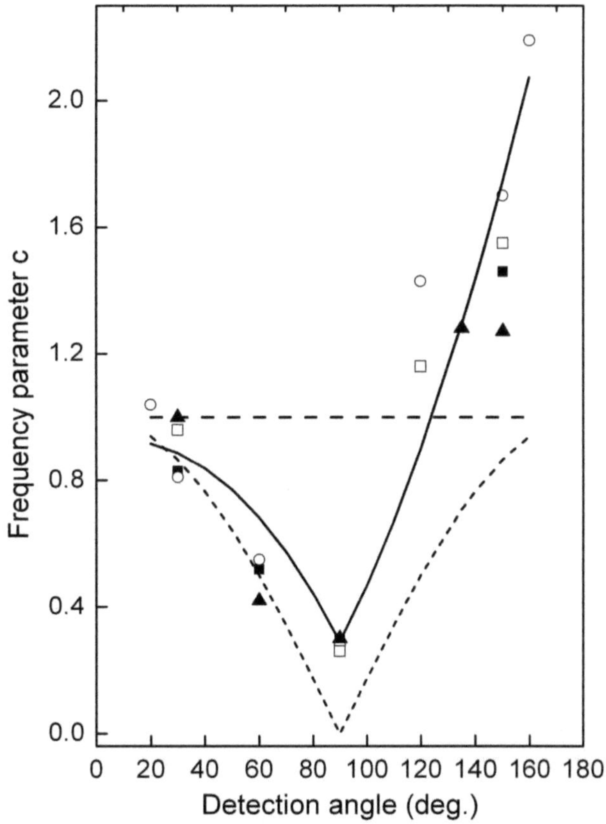

Fig. 3.57 Frequency parameter *c* as a function of the detection angle. Experiment: *full squares*, $Kr^{34+} + H_2$ [19]; *open circles*, $F^{9+} + H_2$ [20]; *open squares*, $H^+ + H_2$ [23]; *full triangles*, 8 keV $e^- + H_2$. Calculation: *short dashed curve* [25]. The *full curve* is the result of a fit of experimental data. The *dashed line* represents the result for $c = 1$

3.4.2.4 The Frequency Parameter

As mentioned in the paragraphs above, the frequency of the observed oscillations is not symmetrical with respect to 90°. More precisely, the frequency at backward angles is larger than that observed at forward angles, whatever the projectile used in the experiment (Fig. 3.57). This is the reason why the oscillation term can be written as $\sin ckd/kd$, where the adjustable frequency parameter c depends on the detection angle. As shown in a previous study [31], the quantity c can be fitted with the function $a + b\cos\theta\left(1 + pe^{\theta}\right)$ (Full curve in Fig. 3.57).

At present, no theory is able to explain this strong asymmetry. Asymmetry is only found for calculations using CDW approximation. However, even in this case, the asymmetry is low compared with the experimental results.

One has to be careful with this parameter c. In fact, the frequency parameter is only defined in a small range of electron velocity. When taking into account a larger range, the mean c value strongly changes, especially at angles close to 90°.

As seen in Fig. 3.58, where the oscillations are represented as a function of the electron velocity in the range 0–30 a.u., the frequency at 90° (red curve) is far from being zero, except in the range 0–5 a.u., where all the experimental studies were performed.

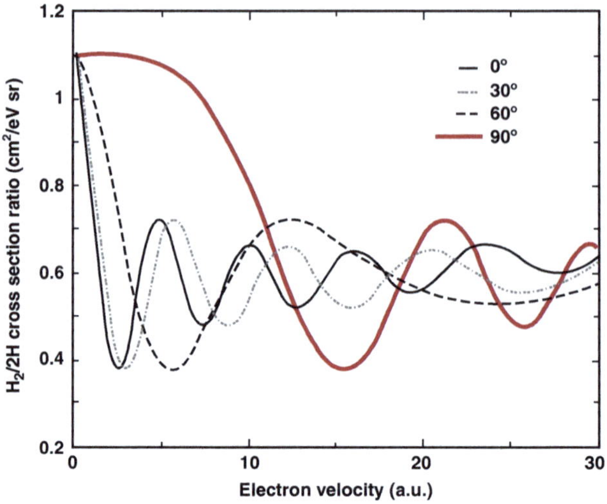

Fig. 3.58 Calculated double differential cross-section ratios for the ionization of H_2 and 2H by 60 MeV/u Kr^{34+} ion impact as a function of ejection angle and electron velocity for various angles. At 90° (*red curve*), the oscillations appear at velocities larger than 10 a.u

3.4.2.5 The Role of Projectile Nucleus in Interferences

Since the interference pattern in the velocity distribution of the emitted electrons is difficult to discern directly, it is very important to find other ways to obtain clear direct evidence for interference. The interference term contains the phase angle, which depends on the molecular orientation, the electron momentum $\overrightarrow{k}$ and the momentum transfer $\overrightarrow{q}$. In all the previous experiments described, the cross section was integrated over $\overrightarrow{q}$, which tends to obscure the phase factor making the interference pattern difficult to detect in the cross sections. For this reason, DDCS for single ionization in $H^+ + H_2$ collisions were measured [32] as a function of the scattered angle θ, for different ejected electron energies.

In Fig. 3.59 the DDCS are plotted (full circles) as a function of θ, for energy losses of 30 and 50 eV. The cross sections decrease as θ increases, reaching a minimum at 1 mrad. A structure appears at larger angles, with a maximum at 1.2 mrad. For a helium target, in contrast (open circles), no structure appears in the DDCS.

Calculations using CDW approximation (dashed curve) are in poor agreement with the experiment, since these calculations cannot reproduce the structure centered at 1.2 a.u. The agreement is much better if the interaction between the projectile and the target is included in the calculations (Fig. 3.59, full curve). This suggests that interference is more sensitive to the scattering angle than to the electron energy.

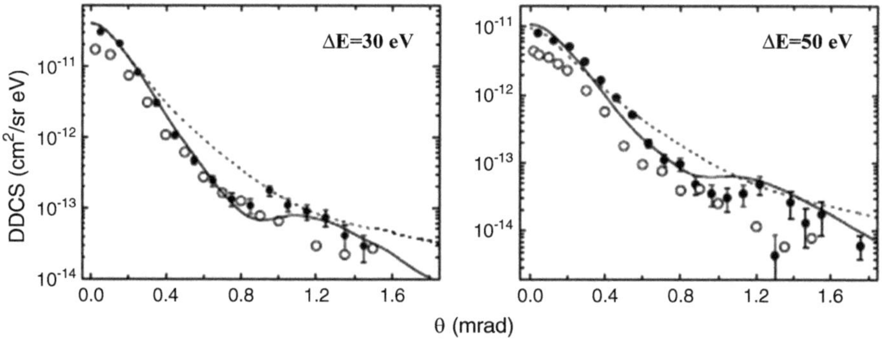

Fig. 3.59 DDCS plotted as a function of the projectile scattering angle for two energy losses (30 and 50 eV). *Full circles*, experiment for H_2; *open circles*, experiment for He; *dashed curves*, CDW calculations; *full curve*, CDW calculation including interaction between the projectile and the target ion

3.4.3 Electron Capture

Electron capture by the projectile can be seen as a particular case of target ionization. In the latter case, the electron can be emitted with any velocity while, for electron capture, the electron is bound to the projectile. The two reactions may be written as follows

$$A^{q+} + B \rightarrow A^{q+} + B^{+} + e^{-} \text{ for ionization}$$
$$A^{q+} + B \rightarrow (A^{q+} + e^{-}) + B^{+} \text{ for capture} \tag{3.18}$$

Because the processes are very similar, electron capture from H_2 by a charged ion is also expected to produce interferences, as in the case of ionization of H_2.

In 2005, two-center interference effects in electron transfer for H^+ + H_2 collisions were investigated at high projectile energies (300–1,300 keV) [33]. In this case, the residual molecular target ion remains mainly bound after the collision

$$H^{+} + H_2 \rightarrow H + H_2^{+}\left({}^{2}\Sigma_g^{+}\right) \tag{3.19}$$

Therefore H_2^+ doesn't dissociate, but recoils with a low kinetic energy. However, the target may also be vibrationally excited, so that the target dissociates, giving rise to a hydrogen atom and a proton, whose energies are of several eV.

$$H^{+} + H_2 \rightarrow H + (H^{+} + H) \tag{3.20}$$

In the experiment, the recoil target ions H^+ or H_2^+ were detected simultaneously with the neutralized projectile. The times and position of the collision partners were recorded. The relative cross sections as a function of the angle between the internuclear H_2 axis and the direction of the projectile are shown in Fig. 3.60.

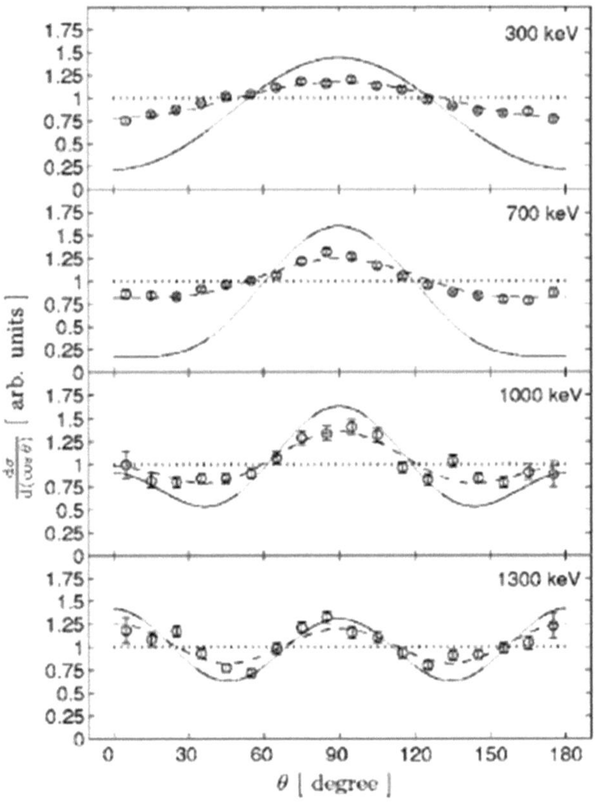

Fig. 3.60 Cross sections for electron capture following $H^+ + H_2$ collisions. Experiment, *open circles*; calculations with capture on the 1s projectile orbital, *full curve*; calculations with capture on several projectile orbitals, *dashed curves*

At all the energies examined, the cross sections are maximal at 90°, where the H_2 axis is perpendicular to the projectile trajectory. At the largest energies, maxima also appear at 0° and 180°, inducing well defined oscillations. If the contribution from the two atomic centers to the transfer probabilities were added incoherently, no angular dependence would appear (dashed curve in Fig. 3.60). Thus, we can infer that all the observed variations of the cross section are due to interferences.

The calculations represented by the full curve show the same qualitative feature and variation with projectile energy as the experimental data. The best agreement is found for the highest energies. At lower energies the agreement is better when the capture is extended not only to the 1s orbital of the projectile, but also for other ℓ states. The result, shown as dashed curves, proves the dramatic influence of such states.

3.4.4 *Ionization of H_2 by Electron Impact*

As detailed in earlier sections, since the interferences reveal the double-centered character of the target, the projectile is intuitively expected to play a minor role in the final interference pattern. However, this is not completely true as interferences may depend on the nature of the massive particles for several reasons. First, differences in the emission spectra corresponding to ion or electron beams appear according to the kinematic conditions imposed by the large difference of projectile masses. For example, binary encounter peaks, where single collisions between the projectile and target electrons are produced, are present at much larger final electron velocities in the ion case. Consequently, interference patterns are expected to be different. Second, for multiply charged ions, the presence of a high projectile charge may considerably modify the profile of the low-energy emitted electrons. Finally, in the case of ion impact, the projectile can be treated classically, even at low energies like a few keV, since the corresponding de Broglie wavelength is small compared to the target dimensions. The situation is completely different for electron projectiles, because their quantum-mechanical behavior has to be taken into account for energies smaller than a few tens of keV.

The first study of interferences using electron projectiles was performed theoretically in 2003 [34]. To compute differential cross sections for ionization of H_2, a molecular three-continuum approximation was employed. Supposing an emitted electron of momentum $\overrightarrow{k}$ in a solid angle Ω, a projectile scattering into a solid angle Ω_s, and protons emitted in a solid angle Ω_p, the fivefold differential cross section can be written

$$\frac{d^5\sigma}{d\Omega\, d\Omega_s\, d\Omega_s\, dE} \propto 2\left(1 + \cos \overrightarrow{\chi}.\overrightarrow{d}\right) \tag{3.21}$$

where $\overrightarrow{\chi} = \overrightarrow{k} - \overrightarrow{\Delta K}$ and $\overrightarrow{\Delta K}$ is the momentum transferred to the ionized electron, and d is the internuclear distance between the protons at equilibrium. When averaging over all the possible molecular orientations, the interference term remains, as in the case of ion projectiles.

Calculations were performed at an incident energy of 4.087 keV, and electron energy of 20 eV, for some fixed molecular orientations. Figure 3.61 shows a typical polar plot obtained at 0°, where the molecule is aligned with the incident projectile direction. The dashed curve is the result of the calculation using two effective H atoms. Two structures appear: the binary peak, located around the momentum transfer direction, and the recoil peak, located in the opposite direction. When the interference term (dotted curve) is included, the initial profile is strongly modified (full curve). Maxima and minima in the intensity distribution, caused by interferences, are clearly identified.

From fivefold cross sections, triple-differential cross sections (TDCS) could be deduced. However, even oscillations are expected after averaging over all possible

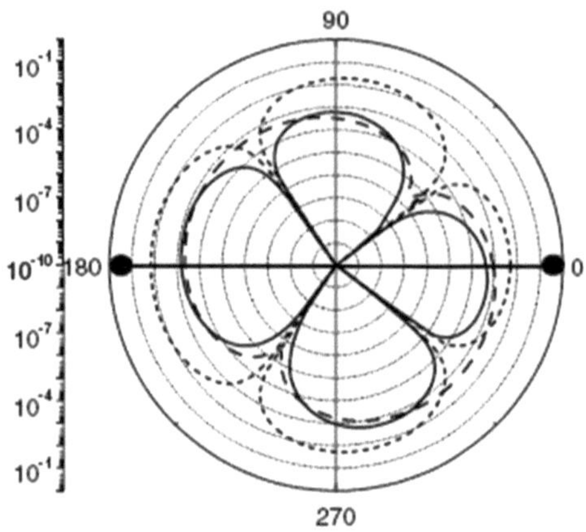

Fig. 3.61 Fivefold differential cross section as a function of the detection angle, for H_2 molecules aligned parallel to the incident beam. The scattering angle is 0°. The *dashed curve* is the result of the calculation without any interference, and the *full curve* takes into account the interferences. The *dotted curve* is the result of the interference itself

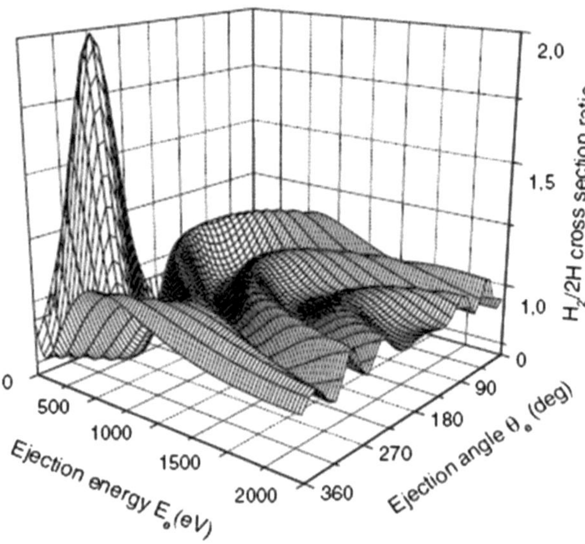

Fig. 3.62 Ratios of molecular TDCS and atom TDCS as a function of the electron energy and the detection angle. The scattering angle is 1°

molecular orientations, the interference effects cannot be clearly seen directly because molecular spectra decrease strongly, by several orders of magnitude, as emitted electron energy increases. Consequently, theoretical ratios between molecular TDCS and twice those corresponding to effective H atoms were determined, and are represented in Fig. 3.62, as a function of the electron energy and the detection angle. The oscillations are well marked, as in the case of ion projectiles.

Experimentally, the first evidence for interference effects was obtained for 2.4 keV electron projectiles on a D_2 target [35]. In Fig. 3.63, typical measured electron spectra are shown as a function of the emitted electron energy (open circles), at detection angles of 30° and 90°. At 30°, in addition to the continuously decreasing cross section, a structure appears, centered at 1.85 keV, caused by binary collisions in which all the momentum is transferred to the electron.

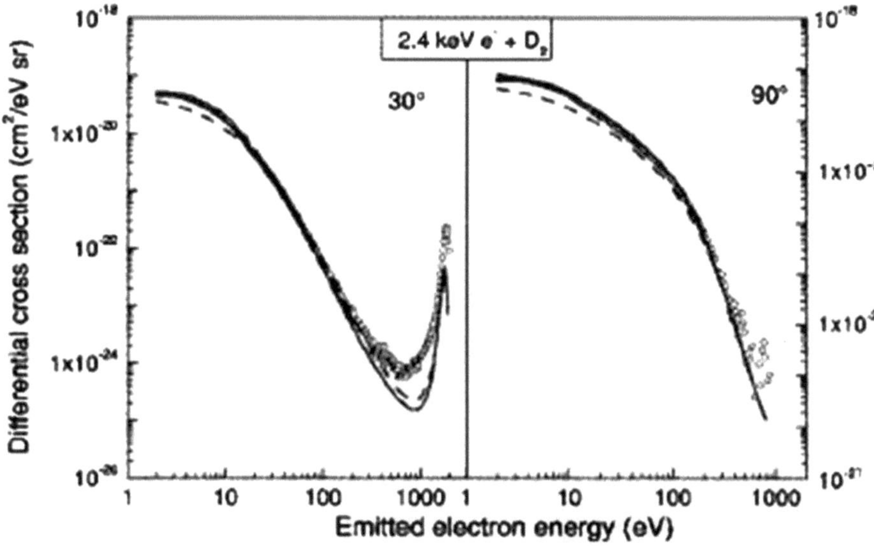

Fig. 3.63 Cross sections for electron emission in 2.4 keV e^- + D_2 collisions as a function of the emitted electron energy, for detection angles of 30° (*left-hand side*) and 90° (*right-hand side*). Experimental data, *open circles*; doubled theoretical cross sections for two effective H atoms, *dashed curves*; theoretical molecular cross sections, full curves

Theoretical results are also shown (Fig. 3.63). The full curves are the result of the two-effective center calculation that includes the interferences, while the dashed curves describe ionization from two independent hydrogen atoms with effective charges. The full calculation including interferences is in very good agreement with the experimental results, except at electron energies above 300 eV.

To make the presence of possible interference more noticeable, as in the case of ion projectiles impacting on H_2, the experimental cross sections were divided by the theoretical atomic cross sections. The so-called experimental ratios are presented in Fig. 3.64 (open circles) for 30° and 70°, and compared with theoretical ratios (full curves).

The overall agreement between experiment and theory confirms the presence of interference patterns due to coherent emission. In contrast with the results obtained with ion projectiles, the frequencies predicted by the theory are close to those of the fitted ratios (dashed curves), at all detection angles.

The behavior of the ratio at 70° should be highlighted. While the theoretical ratio oscillates, the experimental ratio remains constant in the velocity range 2–5 a.u. In addition, the theoretical and experimental ratios oscillate around ~1.7 and not around 1 as in all the other angles. A detailed analysis shows that this enhancement is due to an overlapping of the structures relative to constructive interferences at small ejection velocities and in the binary-encounter region (high velocities).

The enhancement of the ratio due to constructive interferences in the binary region was analyzed in detail for 8 keV e^- + H_2 collisions in 2008 [36]. The

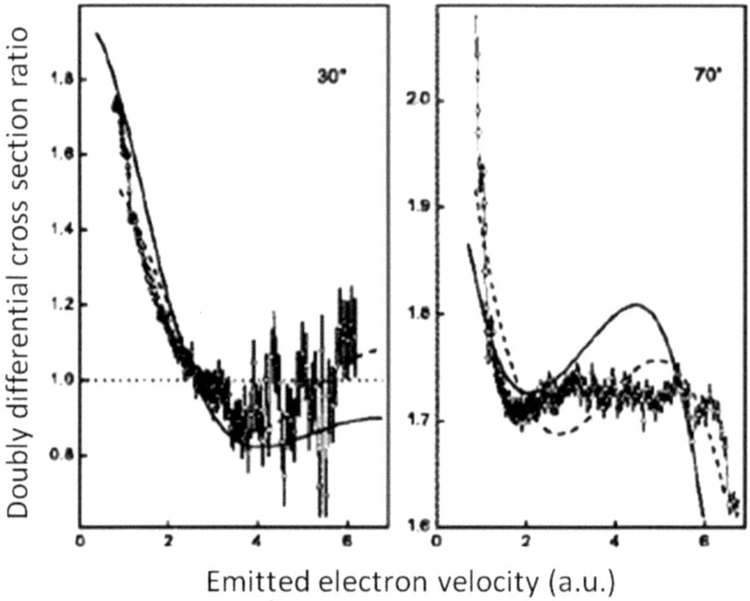

Fig. 3.64 Experimental (*open circles*) and theoretical (*full curves*) doubly differential cross section ratios for electron emission in 2.4 keV e^{-} + D_2 collisions as a function of the emitted electron velocity, for detection angles of 30° and 70°. Experimental ratios are fitted by using a zero-order Bessel function (*dashed lines*)

electron spectra in the binary region were measured and compared with calculations. The angular distributions of emitted electrons with energies of 9, 70 and 280 eV are plotted as a function of the emission angle on the left-hand side of Fig. 3.65. For the three energies, the atomic cross sections underestimate the experimental data by at least 30 % in the binary region around 90°.

To quantify the effects of the interference more precisely, the experimental-to-theoretical ratios were extracted (right-hand side of Fig. 3.65) and compared to theoretical ratios. The ratio is maximum in the binary region, especially at 70 eV, showing the importance of coherence in this domain, as shown previously for 2.4 keV e^{-} + D_2 collisions [34].

The interference effects in electron collisions have been discussed not only according to electron energy distributions, but also from angular distributions. For example, recently, relative TDCS have been measured for the ionization of He and H_2 by impact of ~700 eV electrons [37]. The angular distributions (Fig. 3.66) mainly consist of two structures.

The highest intensity structures, centered at angles in the range 40–60°, and are produced by binary collisions, while the smallest intensity, centered at ~200–240°, are due to soft collisions, giving rise to the so-called recoil electrons. In the case of He (left side of Fig. 3.66), the calculated results (dashed curves), based on convergent close coupling (CCC) calculations, reproduce the experiment very well. The

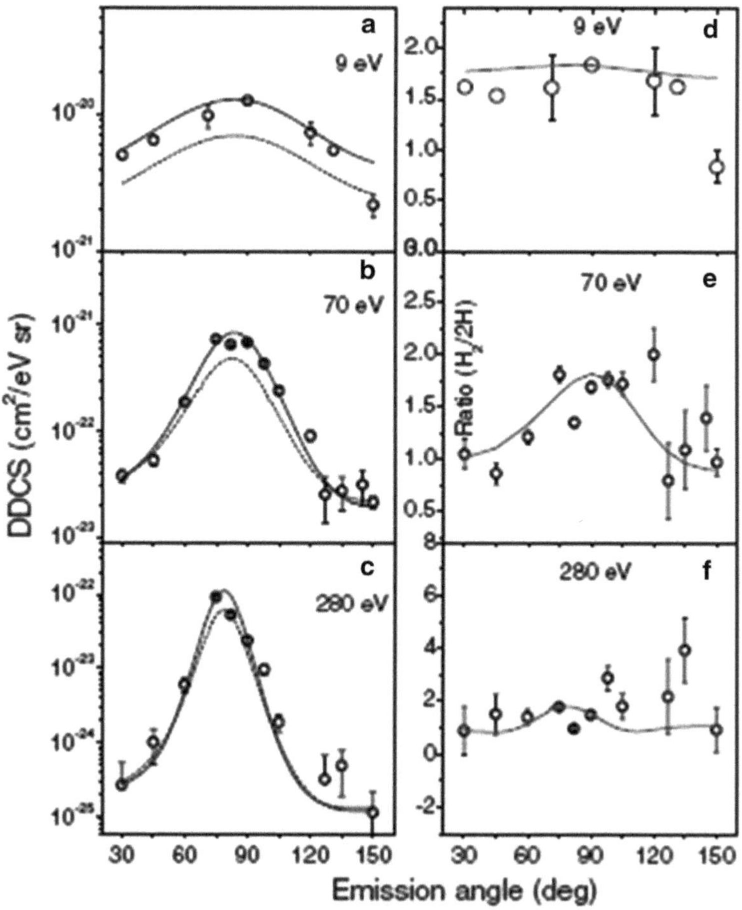

Fig. 3.65 *Left-hand side*: angular distribution of emitted electrons with energies of 9, 70 and 280 eV (*open circles*). Full curve: theoretical molecular cross section. *Dashed curves*: double the theoretical cross section for H atoms. *Right-hand side*: experimental-to-theoretical DDCS ratios (*open circles*). *Full curves*: theoretical ratios

small deviations in the recoil region are partly due to lack of statistics, since the count rates are small. For H_2, two calculations were applied. The first model is based on the first Born approximation, in which the two-center continuum (TCC) approximation is applied [33]. The second is the molecular three-body distorted wave (M3DW) model. The comparison between experiment and theory (right-hand side of Fig. 3.66) is less satisfactory than that for He. The shape of the binary structures is rather well reproduced, even though the calculations shift away from each other. However, both models predict too small a recoil intensity, except for TCC calculations at 74 eV.

The angular distributions for He and H_2 look very similar. A more detailed inspection of the spectra shows that the recoil peak for H_2 is much smaller than that

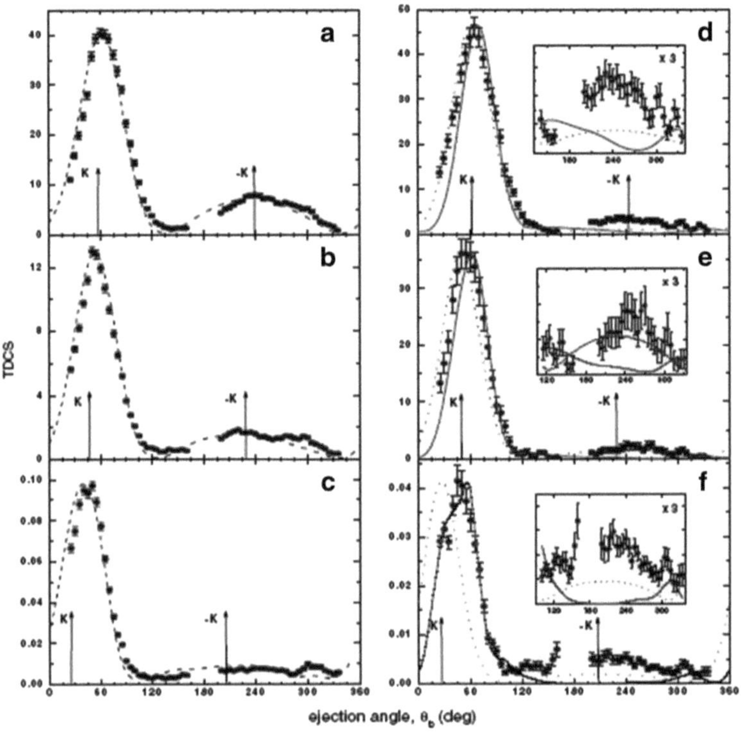

Fig. 3.66 (e,2e) TDCS for ionization of He (*left-hand column*) and H_2 (*right-hand column*), plotted as a function of the ejection angle, at a fixed scattering angle of $-6°$ and a fixed scattering energy, of 500 eV. For He, the *dashed line* represents the results of the convergent close coupling (CCC) calculations. For H_2, the *dotted and full lines* represent the theoretical results from the two-center continuum (TCC) and the molecular three-body distorted wave (M3DW) models, respectively. *Solid circles*: experimental data. The *vertical arrows* indicate the momentum transfer direction and their opposite. The insets in the H_2 results represent a zoom on the low intensity recoil region to facilitate comparison

for He at energies of 35 and 74 eV, and larger at 205 eV. This effect, due to interferences, could be emphasized by dividing the cross sections for H_2 by those for He (full circles in Fig. 3.67).

The ratios are compared to the interference factor, defined as the ratio between the calculated cross section for H_2 and twice that for hydrogen (full curve in Fig. 3.68). For the three electron energies, the agreement between experiment and calculation is rather good, since they give rise to the same angular shape. This suggests that the present observations are really due to destructive or constructive interference effects caused by the two-center character of H_2.

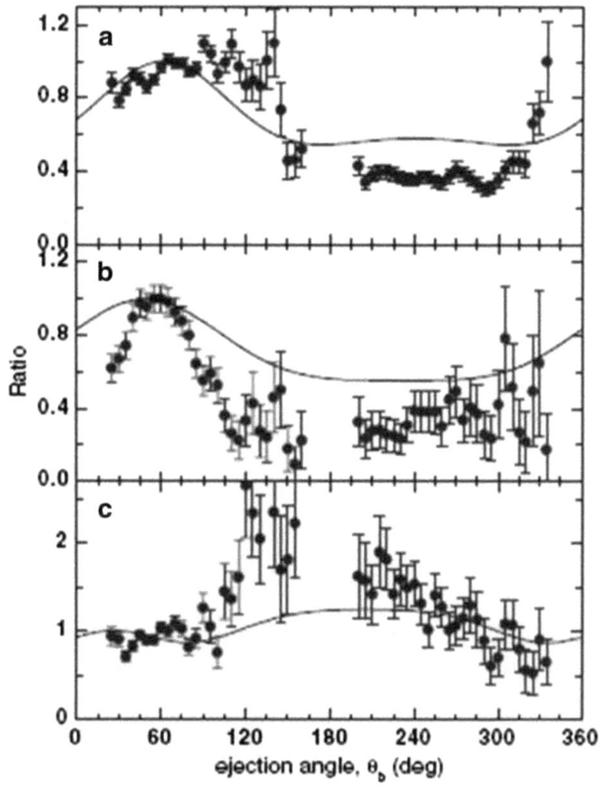

Fig. 3.67 Experimental ratio, $\sigma(H_2)/\sigma(He)$ as a function of the emission ejection angle at 37 eV (**a**); 74 eV (**b**) and 205 eV (**c**). The *full lines* represent the calculated interference factor

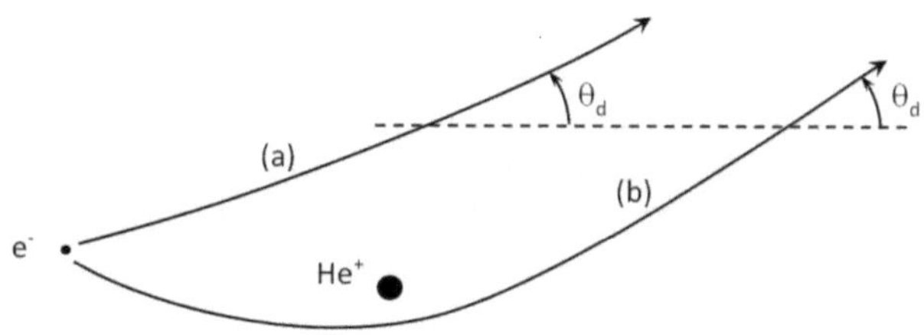

Fig. 3.68 Two different electron trajectories for which electrons are detected with a given energy at the same θ_d angle

3.5 Electron Emission in Atomic Collisions: Young 1804

3.5.1 Principle of the Experiment

In 1991, an interference effect, which was named the Coulomb path interference effect by the authors [38], was demonstrated experimentally and described theoretically for ion-atom collisions. The collision system $He^+ + He$ was investigated at a projectile energy of 10 keV. At this energy, the projectile may capture one target electron or excite the target. In the case of double excitation, the reaction can be written

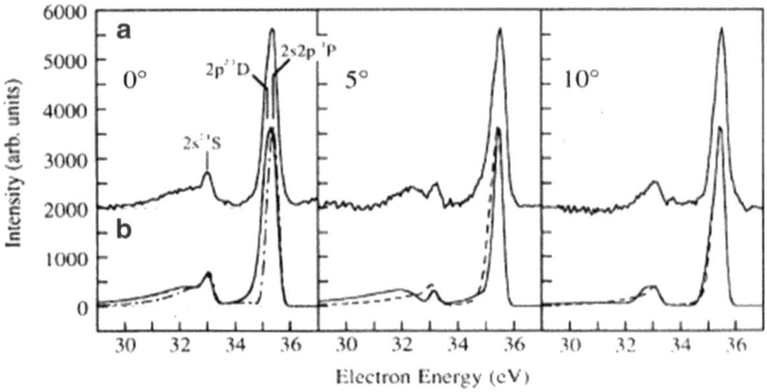

Fig. 3.69 Experimental (**a**) and calculated (**b**) electron spectra for 10 keV He^+ + He at emission angles of 0°, 5° and 10°. Theory: *full curve*, calculation using Coulomb path-interferences; *dashed curve*, calculation where scattering around the "near side" of the projectile is considered; *dot-dashed curve*, calculation at 0° using the OR model

$$He^+ + He \rightarrow He^+ + He^{**}(2\ell 2\ell') \tag{3.22}$$

where $2\ell 2\ell'$ refer to the most dominant doubly excited orbitals. Since these orbitals mainly become deexcited by Auger effect, the post-collision reaction can be written

$$He^{**}(2\ell 2\ell') \rightarrow He^+(1s) + e_A^- \tag{3.23}$$

where e_A^- is the Auger electron.

In principle, the Auger electron is emitted in any direction. If the emission takes place in the vicinity of the projectile, at least two different paths, corresponding to slightly different emission times, may lead to electrons that are detected with an energy E at the same detection angle θ_d. Thus, due to the indiscernibility of the paths, interference may occur, inducing changes in the electron profiles. This effect is similar to that described by Young in his experiment in 1804 (cf. Ref. [21] of Chap. 1), where a pattern of fringes appeared, due to the combination of light waves passing on both sides of a slip of card.

Figure 3.68 shows a schematic view of these different paths. In path (a), the electron is emitted without being deflected by the projectile, while the electron in path (b) is deflected due to the presence of the projectile.

3.5.2 *Interpretation of the Spectra*

Three typical spectra are presented in Fig. 3.69a for the three detection angles of 0°, 5° and 10°, with respect to the incident beam direction. In addition to the main peaks associated with the deexcitation of $2s^2\ ^1S$, $2p^2\ ^1D$ and $2s2p\ ^1P$ configurations,

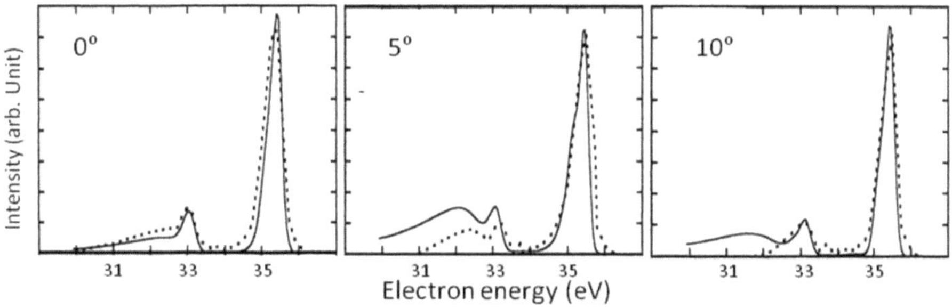

Fig. 3.70 Full curves, calculated spectra for 10 keV He^+ + He at emission angles of 0°, 5° and 10°. *Dashed curves*, experimental spectra [39]

located at 33.3, 35.3 and 35.5 eV, respectively, pronounced shoulders in the $2s^2\ {}^1S$ line shape are observed. At 5°, these become nearly equal in intensity to the main part of the peak.

To determine the theoretical profile of Auger electrons (Fig. 3.69b), the overlapping resonances (OR) model was used [39]. Due to the overlap of the three structures, caused by post-collision interaction between the emitted electron and the projectile (Appendix 4), the resulting line shape was calculated in terms of a coherent sum of amplitudes for emission from each of the resonances (see Sect. 3.6). However, this model, which has no explicit angular dependence, is unable to correctly reproduce the experiment (dotted curve at 0°). The OR model was thus improved. First the Coulomb focusing effect [40] was taken into account. This effect results from the deflection of the ejected electron in the Coulomb field of the projectile. Then, semi-classical phases were included, and the coherent sum over the different paths was incorporated.

The full curve in Fig. 3.69b represents the result of the complete calculation. The model reproduces the experimental profiles well at the three angles considered here. When the path-interference mechanism is removed, the shoulders can be seen to disappear (dot-dashed curve). According to the authors, this proves the important role of path interferences, also known as near-side far-side interferences.

A few years later, the He^+ + He system was revisited theoretically [41]. In this new model, Stark effect was included in the emission angle-dependent autoionization amplitude. Stark mixing has its greatest effect on the tail of the line shape, changes the state population and affects the phase of the autoionization amplitude. Hence, the new model included Coulomb focusing, interferences between the individual structures, and Stark mixing.

Figure 3.70 shows a comparison between experimental spectra from [39] (dashed curves) and spectra calculated using this model (full curves), at angles of 0°, 5° and 10°.

The agreement is reasonable, especially at 0°. At 5°, a small shift in the secondary structure is observed. At 10°, the calculation again presents a shoulder at ~31.5 eV, while no shoulder appears in the experiment. As suggested by the authors, this disagreement may originate from the subtraction of the experimental background due to direct ionization.

To conclude, one sees, through the results of the present calculations, that it is not necessary to invoke Coulomb-path interferences to explain the prominent shoulders observed at small detection angles. The latter effect may exist, but has not proved to play a role as important as previously suggested [39].

3.6 Other Sources of Interferences

The interference phenomenon studied here (Young-type interferences) is not the only one that exists. Unfortunately, the effects of Young-type interferences can be masked by other interferences. As already mentioned in Sects 3.4.2.3 and 3.5, interferences between direct ionization (DI) and autoionization (AI), and interferences between Auger lines themselves, may considerably change the final electron energy profile.

3.6.1 Interferences Between Direct Ionization and Autoionization

Many studies have been performed, experimentally and theoretically, on the role of direct ionization and autoionization processes. One of the most successful studies is undoubtedly the one that was conducted around 20 years ago, in which the post-collision interaction in H^+ + He collision was studied [42]. In this collision system, two effects contribute to the electron profile: the post-collision interaction due to the vicinity of the projectile when the electron is emitted, and the interferences between DI and AI, giving rise to Fano or Shore profiles.

In fact, the authors mention the possible interference between the autoionization profiles themselves. However, this effect has not been taken into account because according to the authors, even with the high resolution used in the experiment, this effect could not be identified.

Figure 3.71 shows an angular dependence of $2\ell2\ell'$ lines shapes for a projectile energy of 36 keV. During the collision, due to the high velocity of the projectile, the H^+ ion excites the target in $2\ell2\ell'$ doubly excited states. At angles ranging from 60° up to 160°, the profile can be well represented using Gaussian curves. This is no longer possible at forward angles. It is clearly seen that, especially at angles smaller than 10°, PCI effects and interferences between DI and AI profoundly change the electron profile. In particular, due to interferences, oscillations with large amplitude are observed for the three configurations involved. The fit reproduces the experimental spectra well in general, except at small angles, where some discrepancies in the amplitude are observed. Finally, it can be noted that the fitting procedure is necessary to extract excitation cross sections. In other words, double excitation cross sections cannot be determined only by integration in energy of the lines.

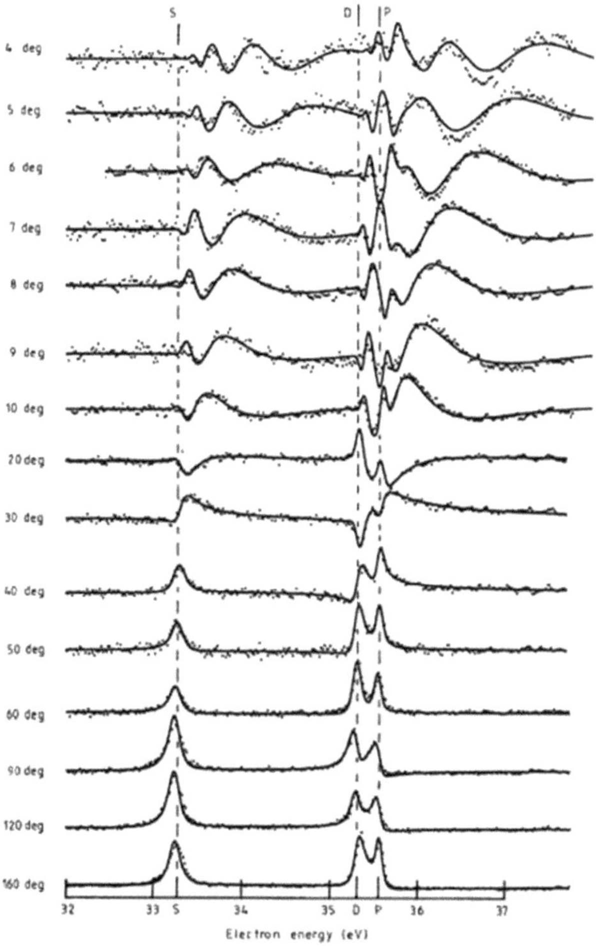

Fig. 3.71 Auger profiles following 36 keV H^+ + He collisions at observation angles ranging from 4° up to 160°

The electron profiles depend heavily on the projectile energy, as shown in Fig. 3.72. At an observation of 4°, for which the PCI effects and interferences are strong, the effects are dominant for large projectile energies. Nevertheless, even at the lowest projectile energies, some additional structures appear, due to PCI, but also to Coulomb path interference, as shown previously [39].

To a lesser extent, interferences between DI and AI can be extracted in collisions where a D_2 target is involved. Double excitation by fast electron impact was studied by measuring electron spectra originating from ionization [43]. After subtraction of the DI contribution, three structures could be distinguished (left-hand side of Fig. 3.73). As seen on the right-hand side, complete calculations were able to reproduce the experiment. While the two structures centered at ~2 and 8.5 eV are due to the Auger deexcitation of doubly excited states (labeled Q_1 and Q_2 on the

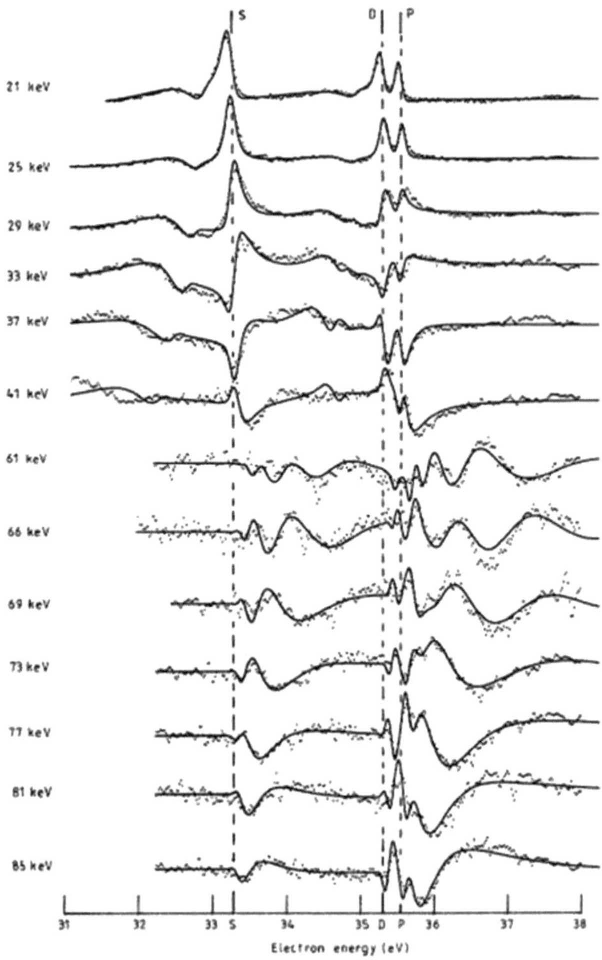

Fig. 3.72 Auger profiles following H^+ + He collisions at an observation angle of 4°, for projectile energies ranging from 21 keV up to 85 keV

Figure), the origin of structure c is more subtle. This structure, which is found to oscillate, results from interference between DI and AI, while the molecule dissociates. As in the case of H^+ + He collisions discussed above, the presence of this structure does not offer the possibility to determine double excitation cross sections, since the interference can be destructive or constructive.

3.6.2 Interferences Between Auger Lines

In Auger structures, the structure overlapping causes interferences between the Auger lines. This phenomenon has been analyzed in detail in the collision He^+ + He at 1.4 keV [39], for example. The Auger spectrum (Fig. 3.74) was measured at 180°,

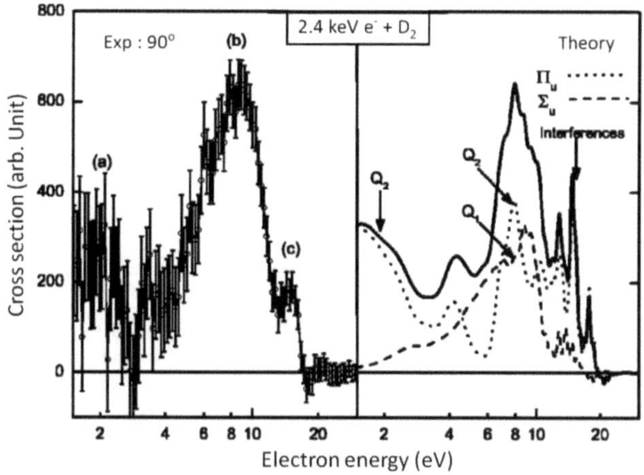

Fig. 3.73 Experimental (*left-hand side*) and theoretical (*right-hand side*) cross sections for Auger electron emission in 2.4 keV e^- + D_2 collisions, as a function of the electron energy. The *error bars* for experimental data originate from statistical standard error

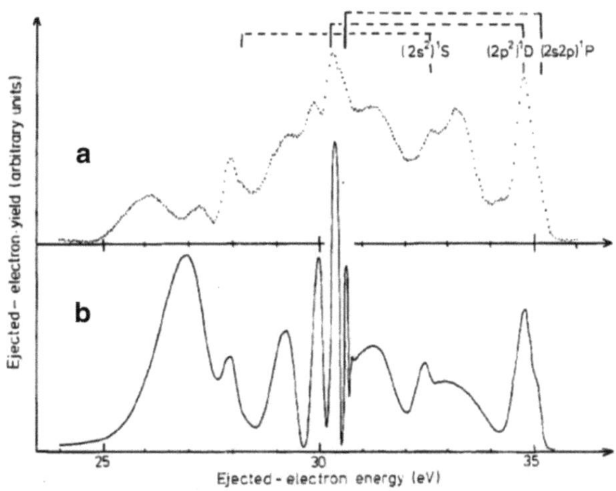

Fig. 3.74 Electron energy spectra obtained in 1.4 keV He^+ + He collisions. Experiment, *top* of the figure; theory, *bottom* of the figure

with respect to the incident beam direction, in order to minimize broadening effects. The spectra consist of three dominant structures, but the interference between AI and DI, as well as interference between the Auger lines, give rise to other complicated structures, especially at the lowest electron energy. The calculation (bottom of Fig. 3.74) reproduces the main features of the experimental data well. The main deviations of the calculated spectrum from the measured one are due to the fact that the small contribution of the $2p^2$ ^1S is was omitted.

References

1. Jönsson, C.: Interferenz von Elektronen am Doppelspalt. Z. Phys. **161,** 454 (1961) Translation by Brandt, D., Hirshi, S.:, Electron diffraction at multiple slits. Am. J. of Phys. **42,** 4 (1974)
2. Feynman, R., Leighton, R., Sands, M.: The Feynman lectures on physics. Addison-Wesley, Redwood City (1964)
3. Merli, P.G., et al.: On the statistical aspect of electron interference phenomena. Am. J. Phys. **44**, 306 (1976)
4. Tonomura, A., et al.: Demonstration of single-electron buildup of an interference pattern. Am. J. Phys. **57**, 117 (1989)
5. Merli, P., et al.: A. Tonomura, Physics World **16,** 20 (2003)
6. Aberth, W., Lorents, D.C.: Elastic differential scattering of He^+ ions by Ne and Ar and of Ar^+ ions by Ar in the 10-600-eV range. Phys. Rev. **144**, 109 (1966)
7. Lockwood, G.J., Everhart, E.: Resonant electron capture in violent proton-hydrogen atom collisions. Phys. Rev. **125**, 567 (1962)
8. Ziemba, F.P., Everhart, E.: Resonant phenomena in large-angle helium ion-helium atom collision. Phys. Rev. Lett. **2**, 299 (1959)
9. Smith, F.T., et al.: Perturbation induced in elastic scattering by crossing of molecular states. Phys. Rev. Lett. **15**, 742 (1965)
10. Itoh, Y.: Measurements of state-selective differential cross section for the one-electron capture in the O^{2+}-He system at $E_{lab} = 30$, 40 and 50eV. J. Phys. B **35**, 3217 (2002)
11. Heil, T.G., Sharma, J.B.: Differential cross sections for the charge transfer reaction O^{2+} + He $\rightarrow$ O^+ + He^+ at low energies. Phys. Rev. A **36**, 3669 (1987)
12. Cohen, H.D., Fano, U.: Interference in the photoionzation of molecules. Phys. Rev. **150**, 30 (1966)
13. Samson, J.A.R., Cairns, R.B.: Total absorption cross section of H_2, N_2 and O_2 in the region 550–200 Å. J. Opt. Soc. Am. **55**, 1035 (1965)
14. Bates, D.R., Öpik, U.: Undulations in the photoionization cross sections curves of molecules. J. Phys. B **1**, 543 (1968)
15. Rolles, D., et al.: Isotope-induced partial delocalization of core electrons in the homonuclear molecule N_2. Nature **437**, 707 (2005)
16. Shafer, N., Bersohn, R.: Photoionization cross section of hydrogenic atoms: a generating function solution. Phys. Rev. A **42**, 1313 (1990)
17. Samson, J.A.R., Haddad, G.N.: Total photoabsorption cross sections of H_2 from 18 to 113 eV. J. Opt. Soc. Am. B **11**, 277 (1994)
18. Fojón, O.A., et al.: Interference effects in H_2 photoionization at high energies. J. Phys. **37**, 3035 (2004)
19. Stolterfoht, N., et al.: Evidence for interference effects in electron emission from H_2 colliding with 60MeV/u Kr^{34+} ions. Phys. Rev. Lett. **87**, 023201 (2001)
20. Misra, D., et al.: Interference effects in electron emission in heavy ion collisions with H2 detected by comparison with the measured electron spectrum from atomic hydrogen. Phys. Rev. Lett. **92**, 153201 (2004)
21. Stolterfoht, N., et al.: Frequency doubling of interference structures in electron emission from H_2 by 68-MeV/u Kr^{33+} + H_2. Phys. Rev. A **69**, 012701 (2004)
22. Misra, D., et al.: Second-order interference in collisions of 4-MeV/u F^{9+} ions with H_2. Phys. Rev. A **80**, 062701 (2009)
23. Hossain, S., et al.: Interference phenomena associated with electron emission from H_2 by (1–5)-MeV H^+ impact. Phys. Rev. A **72**, 010701(R) (2005)
24. Bethe, H.A.: Zur Theorie des Durchgangs schneller Korpuskularstrahlen durch Materie. Ann. Phys. (Leipzig) **5**(325) (1930)
25. Nagy, L., et al.: Interference effects in the ionization of H_2 by fast charged projectiles. J. Phys. B **35**, L453 (2002)

26. Sarkadi, L.: Interference effects in electron emission from H_2 by particle impact. J. Phys. B **36**, 2153 (2003)
27. Galassi, M., et al.: Multicenter character in single-electron emission from H_2 molecules by ion impact. Phys. Rev. A **70**, 032721 (2004)
28. Póra, K., et al.: Interference effects in the differential cross section of H_2 by H^+ impact. Nucl. Instr. Meth. B **233**, 293 (2005)
29. Reading, J.F., et al.: Finite-Hilbert-basis-set calculations for the angular distribution of the ionized electrons produced in p + H impact at 20keV. Phys. Rev. A **70**, 032718 (2004)
30. Kirby, G.W., et al.: Energy and angular distributions of electrons from ion impact on atomic and molecular hydrogen. II. 20-114-keV H^++H. Phys. Rev. A **51**, 2256 (1995)
31. Chatterjee, S., et al.: Electron interference in fast ion collisions with H_2 and the frequency parameter. Phys. Scr. **T144**, 014040 (2011)
32. Alexander, J.S., et al.: Interference effects due to projectile target nucleus scattering in single ionization of H_2 by 75-keV proton impact. Phys. Rev. A **78**, 060701(R) (2008)
33. Stochkel, K., et al.: Two-center interference in fast proton–H_2-electron transfer and excitation processes. Phys. Rev. A **72**, 050703(R) (2005)
34. Stia, C.R., et al.: Interference effects in single ionization of molecular hydrogen by electron impact. J. Phys. B **36**, L257 (2003)
35. Kamalou, O., et al.: Evidence for interference effects in both slow and fast electron emission from D2 by energetic electron impact. Phys. Rev. A **71**, 010702(R) (2005)
36. Chatterjee, S., et al.: Bethe binary-encounter peaks in the double-differential cross sections for high-energy electron-impact ionization of H_2 and He. Phys. Rev. A **82**, 052709 (2010)
37. Staicu Casagrande, E.M.: (e,2e) ionization of helium and the hydrogen molecule: signature of two-centre interference effects. J. Phys. B **41**(025204) (2008)
38. Swenson, J.K., et al.: Coulomb "path" interference in low-energy ion-atom collision. Phys. Rev. Lett. **66**, 417 (1991)
39. Morgenstern, R., et al.: Interference of autoionizing transitions in time-dependent fields. J. Phys. B **10**, 1039 (1977)
40. Swenson, J.K., et al.: Observation of Coulomb focusing of autoionization electrons produced in low energy He^+ + He collisions. Phys. Rev. Lett. **63**, 35 (1989)
41. Cordrey, I.L., Macek, J.H.: Theory of the angle-dependent autoionization cross section in ion-atom collisions. Phys. Rev. A **48**, 1264 (1993)
42. Moretto-Capelle, P., et al.: Distortion of $He2\ell2\ell'$ Fano lineshapes by strong post-collision interaction in H^+ – He collisions. J. Phys. B **29**, 2007 (1996)
43. Chesnel, J.-Y., et al.: Autoionization electrons following double excitation of D2 in 2.4 keV e^-+D_2 collisions: experimental and theoretical evidence. Phys. Rev. A **70**, 010701(R) (2004)

Chapter 4
Young-Type Electron Interferences Using Single Electron Sources

4.1 Principle of a Single-Electron Experiment

4.1.1 Analogy with Photon Experiments

We saw in Chap. 1 (Sect. 1.1.6.2) that a photon beam cannot consist of single particles, even if the beam intensity is strongly attenuated. One can only say that, on average, the number of interacting photons is equal to one. To break through this fundamental problem, it was demonstrated that it is possible to build a single photon source indirectly. A photon beam (Fig. 4.1) strikes a crystal, and excites the atoms. The deexcitation of the atoms induces photons that can be emitted in any direction, but only those that are emitted towards the interferometer produce interferences that can be detected.

Therefore, to ensure the single-electron condition, the electron source has to be created indirectly, by means of a first excitation (right side of Fig. 4.1). In principle, the electrons can be produced either by direct ionization, or by Auger effect. In the case of a direct ionization, the electron can have any energy. In other words, ionization does not give rise to a monochromatic or quasi monochromatic source. The second possibility is thus to produce Auger electrons. Since Auger effect (or autoionization) is a resonant process, the energies are well defined.

4.1.2 An ideal Single-Electron Experiment

In 2004, two theoreticians proposed and described an original experiment [1] in order to observe Young-type double-slit interference patterns using electrons passing one-by-one through a nanoscale interferometer (Fig. 4.2).

The first excitation is provided by a molecular H_2 beam (Fig. 4.2) colliding on fully stripped He^{2+} ions. The He^{2+} ion captures both H_2 electrons in a doubly excited state, namely the $2\ell 2\ell'$ state. Then the He atom becomes deexcited by Auger

F. Frémont, *Young-Type Interferences with Electrons*, Springer Series on Atomic, Optical, and Plasma Physics 77, DOI 10.1007/978-3-642-38479-0_4,

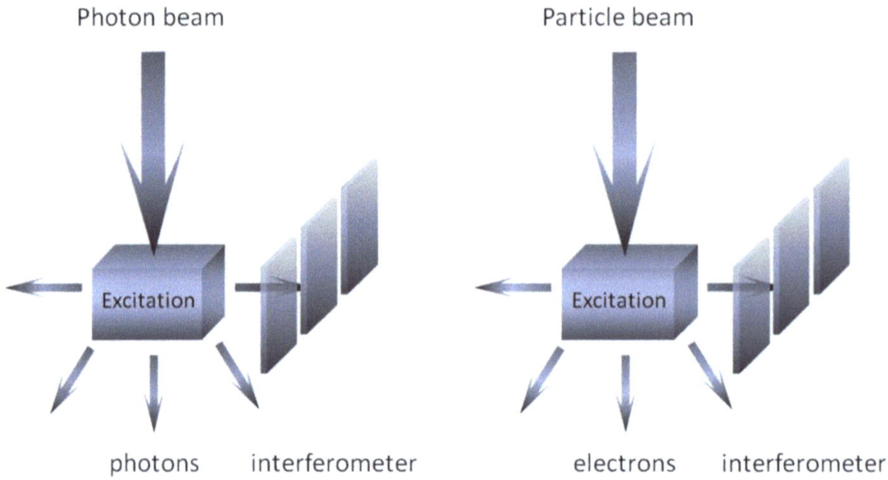

Fig. 4.1 Schematic diagram of the process necessary to produce single photons (*left-hand side*) and single electrons (*right-hand side*). Both originate from a first excitation induced by a photon or particle beam

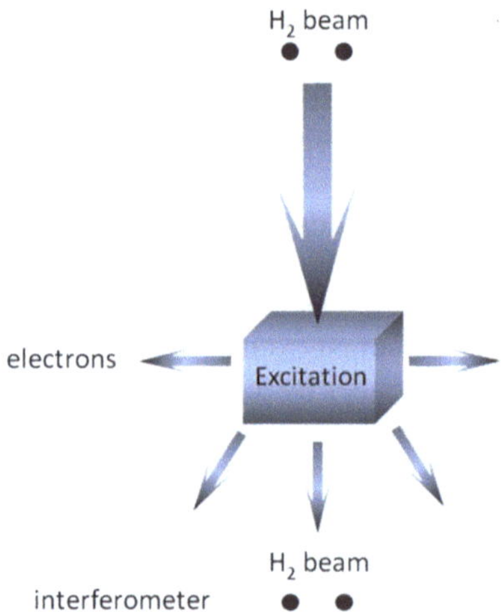

Fig. 4.2 Diagram of a single-electron interference experiment. A He^{2+} ion captures both electrons from H_2 and becomes deexcited by Auger effect. When the electron scatters on both protons, it produces interferences

effect, producing electrons in any direction, whereas H_2^{2+} separates into two protons. The electrons that are emitted in the direction of the protons scatter on them. The protons play the role of the slits, and interferences occur. To summarize, in this experiment, H_2 plays the roles of both the excitatory agent and the interferometer.

In practice, a low energy (<100 keV) He^{2+} ion beam collides with a jet of H_2 that is supposed to be at rest in the laboratory frame. During the collision, several reactions may occur. One obtains the following, for single capture (SC), double capture (DC) and direct ionization (DI)

$$He^{2+} + H_2 \rightarrow He^{+} + {H_2}^{+} \quad (SC) \tag{4.1a}$$

$$He^{2+} + H_2 \rightarrow He + {H_2}^{2+} \quad (DC) \tag{4.1b}$$

$$He^{2+} + H_2 \rightarrow He^{2+} + {H_2}^{+} + e^{-} \quad (SC) \tag{4.1c}$$

In the case of DC, the projectile is mainly in its fundamental state $1s^2\,{}^1S$, or in a $1sn\ell$ singly excited state. There is also a non-negligible probability for the projectile to be in a $2\ell 2\ell'$doubly excited state. After the collision, an Auger electron is emitted with an energy of ~33–35 eV, depending on the excited state involved. When the electron is emitted at backward angles, with respect to the incident beam direction (Fig. 4.3), it scatters on the protons, giving rise to interferences that can be, in principle, detected. Thus, the experiment consists of an ensemble of elementary experiments in which “individual scattering processes are always repeated with similar initial conditions” [1]. In contrast with previous experiments where the electron originated from the interferometer itself, the present experiment is analogous with the Young double-slit experiment at a nanoscopic scale, since the electron source and the interferometer are separated.

4.1.3 Visibility of the Expected Fringes

The interferences created by electrons scattering on H^+ manifest themselves in the production of maxima and minima in intensity, as in the case of photon interferences. The fringe spacing is approximately given by $i = \lambda D/d_{HH}$, where λ is the electron wavelength, D is the distance between the detector and the interferometer, and d_{HH} is the distance between the two protons. Since the detector is located on a circular ring (cf. Sect. 4.2.1), it is relevant to introduce the angular period T defined by tan $T = i/D = \lambda/d_{HH}$. The distance at equilibrium is $d^o_{HH} = 1.42$ a.u., and the wavelength associated with an electron of energy equal to 33 eV is ~0.3 a.u. Thus, the period would be of the order of 80°, which is too large to allow an interference pattern to be observed. However, after the capture, the distance between the protons increases with a velocity $v_{HH} \sim 0.0195$ a.u. Since the autoionizing state has a mean lifetime Γ, an average distance between the protons can be defined. For example, in the case of the $2s^2\,{}^1S$ state, $\Gamma \sim 0.005$ a.u., so that the average deexcitation time is ~2/Γ ~ 400 a.u. Thus, the average distance is $\overline{d_{HH}} \approx d^o_{HH} + v_{HH}\,t \sim 17$ a.u., which is considerably larger than the equilibrium distance. Consequently, the period T strongly decreases. For the $2s^2\,{}^1S$ state, the predicted value of T is 17°, which is well adapted to the experiment.

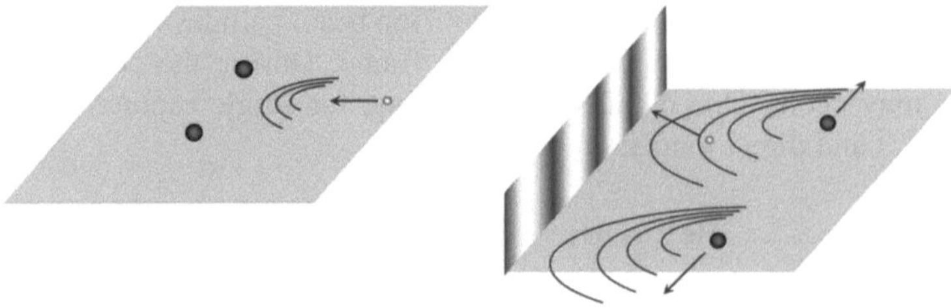

Fig. 4.3 The electron, emitted by Auger effect (*left-hand side*), scatters on both protons (*right-hand side*). The waves that are created interfere after scattering, producing maxima and minima in intensity

4.1.4 Limitations of the Contrast

In the present experiment, the contrast is limited by several parameters. First, the contrast is limited by the width of the total Auger structure. Since many configurations will be involved during the collision, the total width will be of the order of several eV, compared to the total range of Auger energies.

Second, as shown in the previous paragraph, the electron can be emitted at any distances between the projectile and the protons. This has to be weighted by the Auger deexcitation probability, which is proportional to $e^{-\Gamma t}$. Therefore, the experiment consists of many individual experiments in which the distance between the slits (i.e. the protons) is arbitrary.

Figure 4.4 shows three angular dependences showing the effect of time integration. On the left-hand side of the figure, the intensity is calculated at time $t = 400$ a.u. The oscillations are clearly visible, with amplitude that decreases as the detection angle decreases. As the difference between the integration limits increases (center of Fig. 4.4.), the amplitude of the secondary maxima strongly decreases. Finally, for time varying from 0 to infinity (right-hand side of the figure), the oscillations are seen to completely vanish. This problem will be discussed in more detail when quantum mechanics calculations are used to describe autoionization.

In addition, the orientation of the molecule is arbitrary. Statistically, this means that the experiment is similar to a diffraction experiment rather than a double-slit experiment. This is the reason why, as described in Fig. 4.4, the angular distribution is characterized by a Bessel function of order 0, i.e. sin θ/θ instead of a pure sinusoidal function.

The angular distribution has been theoretically analyzed for fixed-in-space molecules (Fig. 4.5) as well as for randomly-oriented molecules (Fig. 4.6).

In the case of fixed-in-space molecules, a first maximum appears at 0°, due to focusing effect, also called Glory effect, and secondary maxima are clearly visible. For randomly oriented molecules, (Fig. 4.6), the shape of the energy and angular distribution is quite similar to the previous case, except that the intensity of the secondary maxima (indicated here by arrows) is much smaller than that for fixed-in-space molecules.

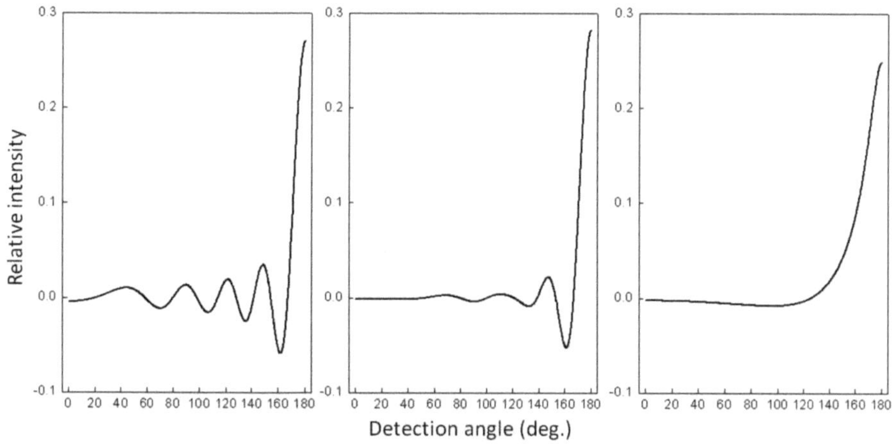

Fig. 4.4 Calculated angular distribution using a Bessel function of 0 order, weighted by Auger deexcitation probability. *Left-hand side*, $t = 400$ a.u.; center, t in the range 300–500 a.u.; *right-hand side*, t in the range from 0 to infinity

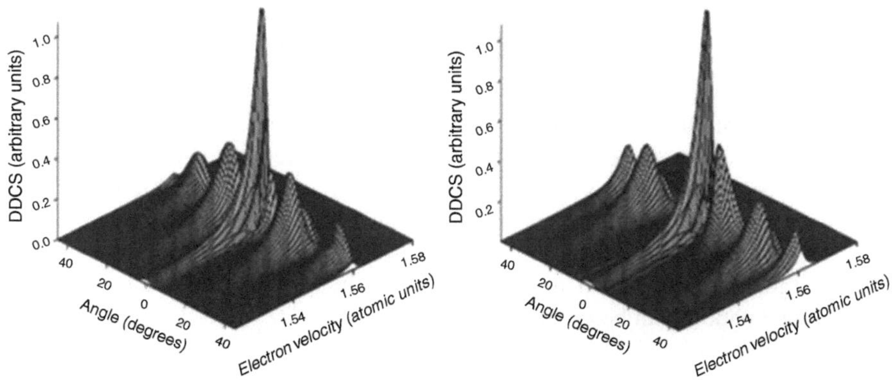

Fig. 4.5 Energy and angular distribution for the system $He^{2+} + H_2$ for two orientations of molecular target

Finally, we should remember that Auger emission is accompanied by electron emission due to direct ionization. This latter process, which can also interfere with autoionization gives rise to a non-negligible background. Therefore, the subtraction of this background, which is necessary to only retain Auger spectra, may lead to a diminution of the contrast

Fig. 4.6 Energy and angular distribution for the system $He^{2+} + H_2$ for randomly oriented molecular targets

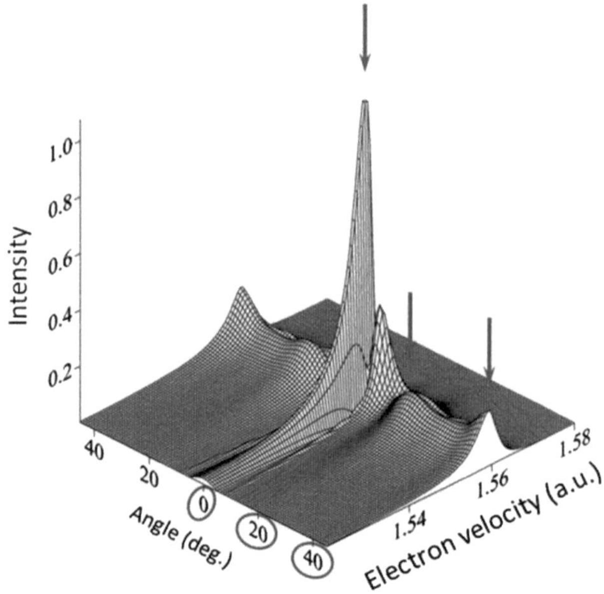

4.1.5 Coherence Time and Coherence Length

As in the case of photons, a coherence time and a coherence length may be defined. The coherence time is defined as the inverse of the frequency width $\Delta\nu$ of the spectrum. Since the energy width is of the order of the mean lifetime Γ, the coherence time is of the order of 200 a.u. Thus, the coherence length $L_c = v_e t_c$, where v_e~1.5 a.u. is the electron velocity, is of the order of 300 a.u. This length is much larger than the mean distance between the protons (~17 a.u.), thus the interference pattern can be observed.

4.2 Realization of the Experiment

4.2.1 One and Two Event Probabilities

4.2.1.1 Interference Condition with One Electron

We will define p_1 as the probability that only one electron scatters on two protons after double capture and Auger effect, and p_2 as the probability that two electrons originating from two different emitting ions scatter on the same interferometer (Fig. 4.7).

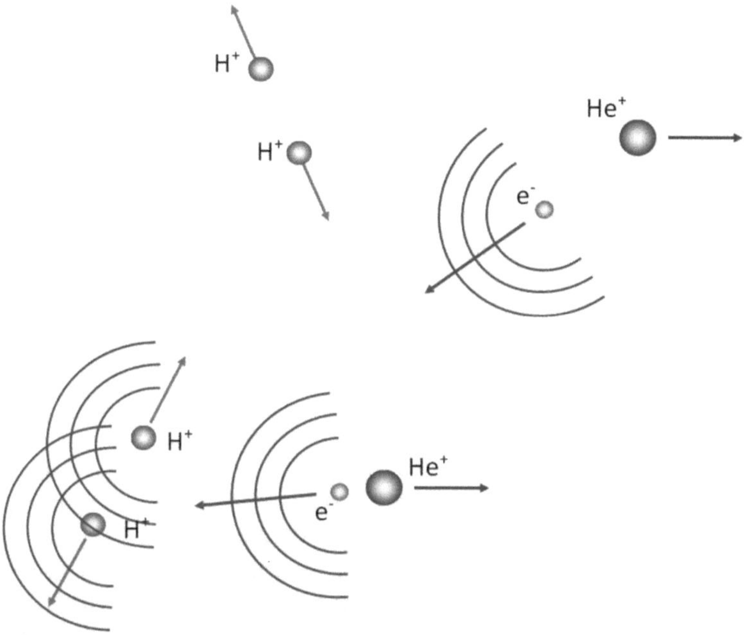

Fig. 4.7 Two independent collisions (*top and bottom*) in which a He projectile captures both electrons from a H_2 molecule and emits an Auger electron. Both Auger electrons scatter on the same interferometer

In a similar way to single photon interferences, the ratio $\alpha = p_2/p_1^2$ has to be smaller than 1 to fulfill the single electron condition. To estimate this ratio, one can see that it is of the same order of magnitude as the ratio of the solid angles considered for each projectile ion (Fig. 4.8), so that

$$\alpha = \ell^2/L^2 \tag{4.2}$$

where ℓ is the distance between the projectile and the protons after double capture and Auger effect in the same event and L is the distance between another projectile that emits an electron after a collision with another target and the same protons (Fig. 4.8).

The determination of α only requires knowledge of the geometric quantities, which mainly depend on the collision parameters: the number of projectile ions per second, the geometry of the beam, and the number of target atoms per volume unit.

4.2.1.2 Collision Parameters and the Single Electron Condition

The number of He^{2+} ions per second is $N_i = I/q$, where $q = 2$ is the projectile charge and I is the beam intensity, of the order of 100 nA; therefore $N_i \sim 3 \times 10^{11}$/s. The DC

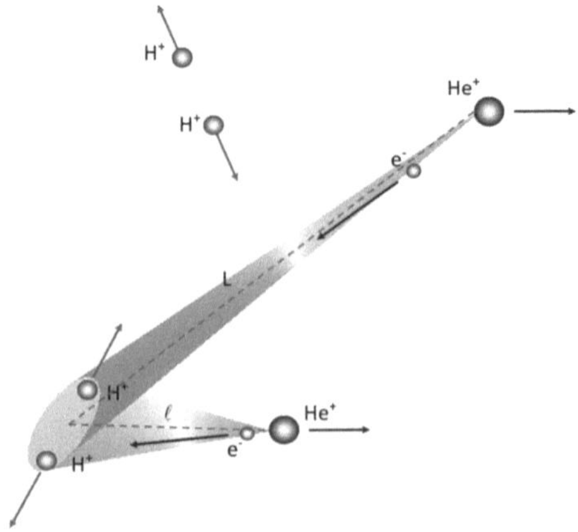

Fig. 4.8 One electron is emitted by Auger effect after a double capture at a distance ℓ from the protons. Another electron, from another projectile, is emitted at a distance L from the same protons

cross section is $\sigma_{DC} \sim 10^{-17}$ cm^2. The probability for a DC event is given by $p_{DC} \sim n_c \ell_c \sigma_{DC}$, where n_c is the target density and $\ell_c \sim 0.4$ cm is the effective collision length. For a typical gas jet pressure of 10^{-4} mbar, $n_c \sim 10^{11}$ cm^{-3}. Thus, $p_{DC} \sim 10^{-6}$, and the number of emitting ions is $N_i\, p_{DC} \sim 3\ 10^5$/s. These ions occupy a volume $V_i \sim \pi R^2 v_p$, where $R \sim 2.5$ mm is the beam radius. In this case, with $v_p = 0.63$ a.u. ~1.2 10^6 cm s^{-1}, $V_i \sim 2\ 10^5$ cm^3. Consequently, the average distance L between two emitters is of the order of 1 cm.

The distance ℓ between one emitter and the protons is $\ell \sim v_p/\Gamma \sim 200$ a.u. ~ 1.3 10^{-6} cm. This gives rise to a ratio $\alpha \sim 10^{-12}$, which is much smaller than 1. The result of the ratio estimation can be interpreted as follows. Suppose 10^{12} events for which only one electron scatters on an interferometer. Among all these events, only one event, in which two electrons originating from two different emitters scatter on the same interferometer, is possible. This means that one would have to wait for approximately three months before this undesire event occurred, which is much larger than the experimental acquisition time (~1 day).

4.2.2 *The Experimental Setup*

The experiment was carried out in 2006 at the low energy ion source facility in Ganil (Grand Accélérateur National d'Ions Lourds), Caen. For reasons that will be explained later, He^{2+} was used at energies of 30 and 8 keV, and N^{6+} at 105 keV. According to this method, the ion beam, delivered by an ECR (Electron Cyclotron Resonance) ion source, crosses an effusive gas jet of H_2 in a scattering chamber (Fig.4.9). The ions are collected in a Faraday cup for the normalization of the cross sections. The emitted electrons, at a fixed detection angle, are analyzed in terms of energy by means of a single-stage spectrometer, which consists of parallel plates. By varying the potential between the plates, a spectrum is obtained, consisting of a number of electrons as a function of their energy.

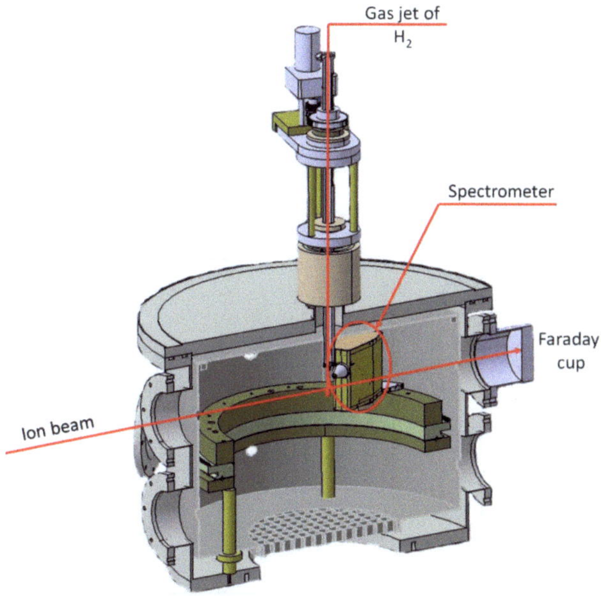

Fig. 4.9 View of the experimental device. The ion beam crosses a jet of H_2 and the projectiles are counted in a Faraday cup. The emitted electrons are analyzed in terms of their energy by using an electrostatic spectrometer

The angular distribution is determined by varying the position of the spectrometer relative to the incident beam direction. The available angular range is $20-160^{\circ}$. For the particular case of He^{2+} + H_2 collisions, a complementary experiment was realized at Centro Atomico Bariloche (Argentina) between 160° and 180°, at a projectile energy of 40 keV, which is the minimum energy available at this ion source.

4.2.3 *Spectral Analysis for 30 keV He^{2+} + H_2 Collisions*

Two typical spectra following 30 keV He^{2+} + H_2 collisions are presented in Fig. 4.10 [2–4] at detection angles θ_d of 160° (left-hand side) and 120° (right-hand side). Two contributions are visible: a monotonous decreasing background, due to direct ionization (DI) of H_2, and well defined structures labeled a, b and c, superimposed onto DI. The three structures are attributed to the Auger deexcitation of He after DC on $2s^2\ {}^1S$, $2p^2\ {}^1D$ and $2s2p\ {}^1P$ and $2\ell n\ell'$ (n > 2) configurations.

To analyze the Auger peaks, the DI contribution was subtracted from the total spectra, after fitting the continuous background with a polynomial function of order 2 in the log-log scale. To compare with theory [1], the $2s^2\ {}^1S$ peak has to be integrated in energy. However, as shown in Fig. 4.10, the width of the peaks increases with decreasing θ_d. This is due to kinematics effects and spectrometer parameters (apertures and resolution). Therefore, the angular distribution was obtained by integration in energy of the total Auger contribution.

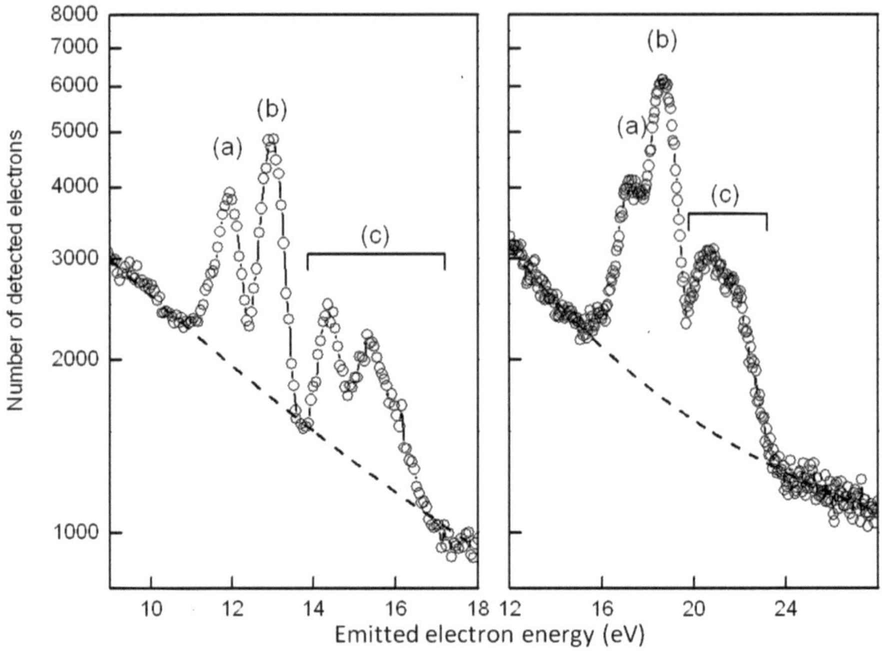

Fig. 4.10 Typical spectra for 30 keV $He^{2+} + H_2$ collisions at 160° (*left-hand side*) and 120° (*right-hand side*). The *dashed line* shows the contribution of DI. The peaks a, b and c originate from Auger deexcitation from the projectile after DC on $2\ell n\ell'$ ($n \geq 2$) configurations

One might ask in what ways this procedure is relevant. Let us consider the three structures involved in the $2\ell 2\ell'$ configurations. The energies of the three peaks are 33.27, 35.31 and 35.55 eV, respectively, associated with mean lifetimes of 5.15×10^{-3}, 2.57×10^{-3} and 1.4×10^{-3} a.u. Therefore, as shown at the top of Fig. 4.11, each of these three peaks, which were obtained using Bessel functions with an angular step of 2°, gives rise to angular distributions with pseudo periods of ~17°, 8° and 4°. In principle, the combination of the three structures may reduce the contrast. However, as seen at the bottom of Fig. 4.11, the sum of the three contributions does not cancel the oscillations. In fact, the oscillations remain, and the pseudo period is very similar to that obtained for the $2s^2\ ^1S$ configuration. The oscillatory behavior of the $2p^2\ ^1D$ and $2s2p\ ^1P$ configurations is lost due to the angular step.

The angular distribution is presented in Fig. 4.12 in the range from 20° to 180°. At forward angles, (open circles), no oscillation is visible. This is not surprising, since the electrons are emitted in the opposite direction to the protons, i.e. in the direction of the projectile. In contrast, clear oscillations appear at angles larger than 90°, i.e. backward angles (full circles and squares at 30 and 40 keV, respectively). In the case of 40 keV $He^{2+} + H_2$ collisions, a maximum is obtained at 180°, and the first secondary maximum is located at about 160°.

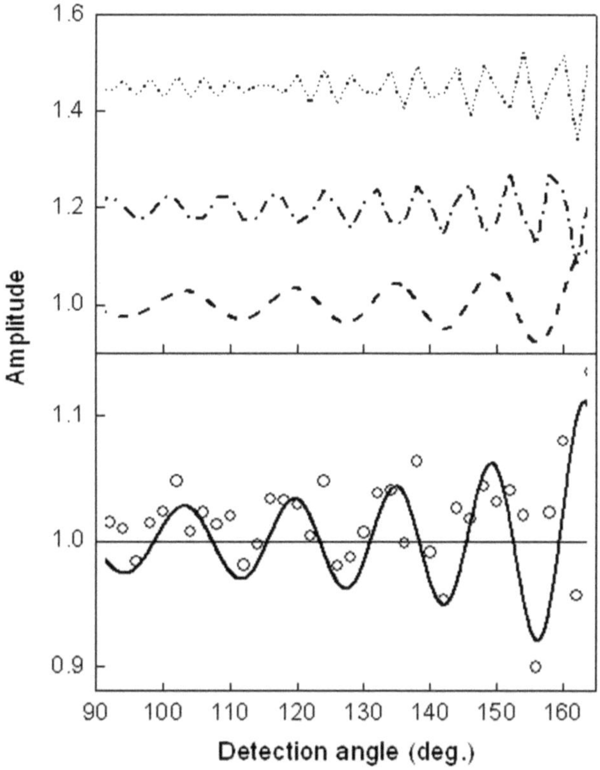

Fig. 4.11 *Top*: Predicted amplitude of oscillations for the $2s^2\ ^1S$ (*dashed curve*), $2p^2\ ^1D$ (*dot dashed curve*) and $2s2p\ ^1P$ (*dotted curve*). *Bottom*: sum of the three contributions (*open circles*). The *full curve* is a fit of the sum

To obtain clear evidence for oscillations and quantitative results, such as the amplitude and the period of the oscillations, the monotonous variation of the angular distribution was subtracted. The result is shown in Fig. 4.12, for angles in the range 90–180°. The amplitude of the oscillations is larger than the uncertainties, especially between 110° and 170°. The relative uncertainties, that take into account the fluctuation of the projectile beam intensity and the variation of the target pressure, are less than 3 %. The pseudo period of these oscillations is of the order of 17°, confirming the predictions of the calculations [1].

To go further with the analysis, a fitting procedure was adopted in order to quantify the different peak contributions. To do this, a path-interference model was utilized (cf. Appendix 5). This model is based on the calculation of the difference between paths traveled by the electron. The calculation shows that the oscillations can be described by the term

$$\sin\delta/\delta \tag{4.3}$$

where

$$\delta = 2\, v_e(\theta_d)\, \overline{d_{HH}} \cos\frac{\theta_d}{2} \tag{4.4}$$

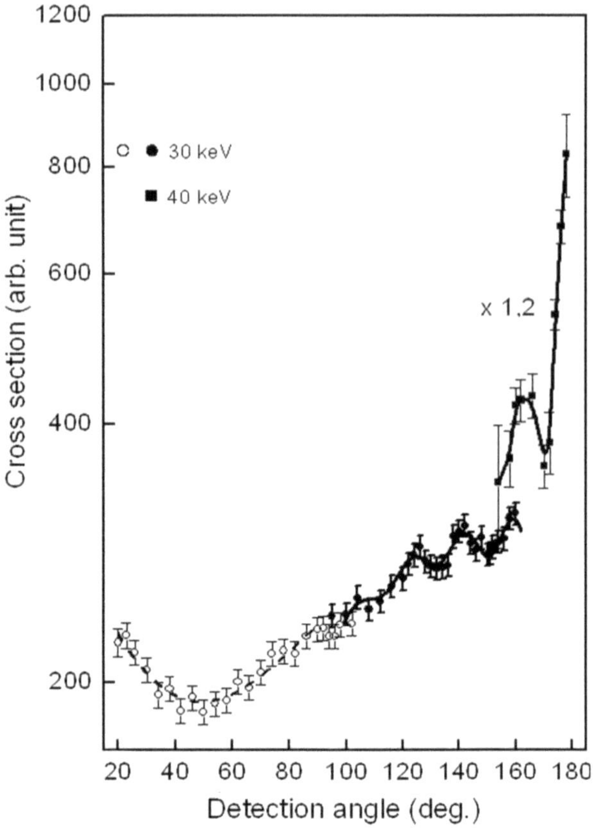

Fig. 4.12 Angular distribution of Auger electrons obtained by integration of the Auger spectra. *Open circles*: forward angles; *full circles* and *squares*, backward angles for 30 and 40 keV collisions, respectively

In the latter expression, $v_e(\theta_d)$ is the electron velocity in the laboratory frame and $\overline{d_{HH}}$ is the average distance associated with the mean time t_A of projectile Auger deexcitation, defined by $t_A = 2/\Gamma$. The term $\sin\delta/\delta$ was first derived by Debye [5] and Ehrenfest [6], who applied it to X-ray diffraction by molecules.

The oscillation amplitude was fitted using a Bessel function of zero order (full curve in Fig. 4.13 for 30 keV and dashed curve for 40 keV). At 30 keV, it is necessary to take into account the three $2\ell 2\ell$' configurations to reproduce the experiment. If only the $2s^2\ ^1S$ configuration is used (Fig. 4.14), it is seen that two sets of values are possible for $\overline{d_{HH}}$ (16 a.u. for the full curve and 20 a.u. for the dashed curve). In contrast, the contribution of the three peaks better reproduces the experiment (Fig. 4.13).

4.2.4 Evolution of Oscillations with Interference Parameters

In photon interference experiments, the shape of the oscillations depends on the interferometer parameters (the distance between the slits and the wavelength).

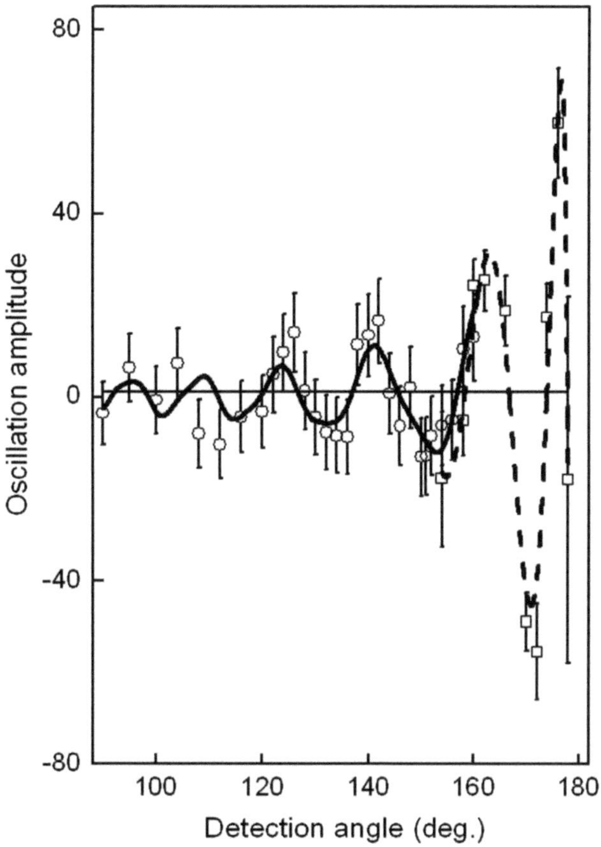

Fig. 4.13 Oscillations obtained from the intensity (Fig. 4.11) by subtracting the monotonous contribution. *Open circles*: experiment; *full* and *dashed curves*: fit using Bessel function for 30 and 40 keV collisions, respectively

Thus, in electron interference experiments following atomic collisions, oscillations are expected to evolve when the values of $\overline{d_{HH}}$ and λ change.

To study the evolution of the interference pattern with λ, the velocity of He^{2+} was divided by a factor of ~2 ($v_p \sim 0.32$ a.u.), corresponding to an energy of 8 keV. Therefore, the electron velocity increases from 0.93 a.u. to 1.32 a.u., and the wavelength is now 0.25 nm, instead of 0.35 nm, at 30 keV. The angular distribution is shown in Fig. 4.15, compared with that obtained at 30 keV [7]. In the angular range explored, the period T is nearly unchanged. However, an angular shift of $T/2$, corresponding to a phase shift of $\pi/2$, is observed. The result of the path interference model (Appendix 5), shown on the right side of Fig. 4.15, is in good agreement with the experimental results.

To change the average distance $\overline{d_{HH}}$, the easiest way is to change the natural width of the peaks. This was done by using N^{6+} as a projectile instead of He^{2+}. The N^{6+} ion captures both H_2 electrons in $2\ell n\ell'$ configurations and becomes deexcited by Auger effect, giving rise to Auger electrons whose energies are of the order of

Fig. 4.14 Experimental cross section (*open circles*) for electron emission in 30 keV He^{2+} + H_2 collisions as a function of the detection angle. The *full and dashed curves* are the result of a fitting procedure using a Bessel function of order 0 with the $2s^2\ {}^1S$ configuration only

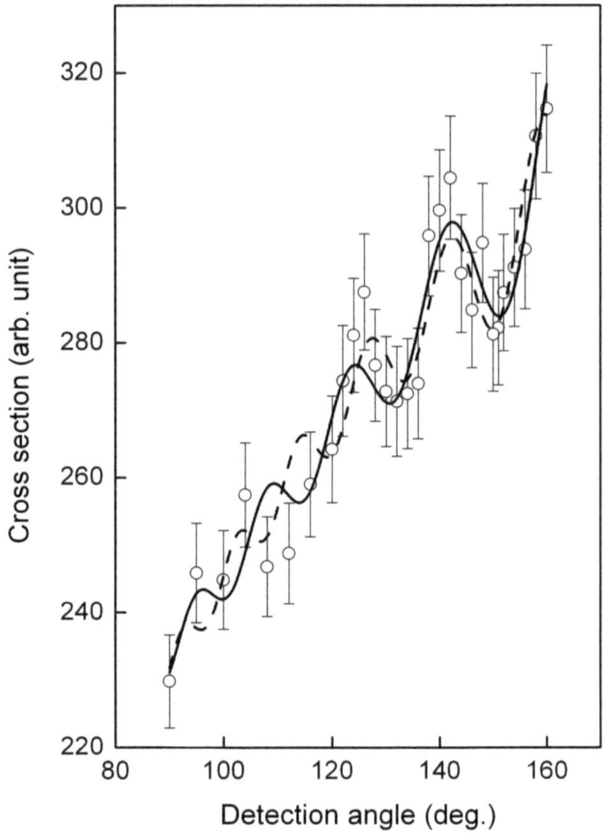

250 eV. The natural width of the peaks associated with $2\ell2\ell'$ configurations is less than 0.001 a.u., which is five times smaller than that for the $2s^2\ {}^1S$ configuration of He. The distance is approximately $\overline{d_{HH}} \approx d^o_{HH} + V_{HH}t$, where $d^o_{HH} = 1.42$ a.u. is the distance at equilibrium, and $V_{HH} \sim 0.0195$ a.u. is the proton velocity.

Therefore, the predicted period is $T \sim \lambda/\ \overline{d_{HH}} \sim 0.6/100 \sim 0.3^\circ$. The period is so small that, due to the resolution of the spectrometer (~2°), no oscillation is visible (Fig. 4.16).

4.3 The Origin of the Oscillations in Electron Interferences

The oscillations, whose amplitude can be understood well by using the Bessel function $\sin\delta/\delta$, depend, at a given angle, on the velocity of the electron, i.e. its energy. Since an Auger peak can be characterized by the intensity and the full width at half maximum (FWHM), the observed oscillations may originate from the

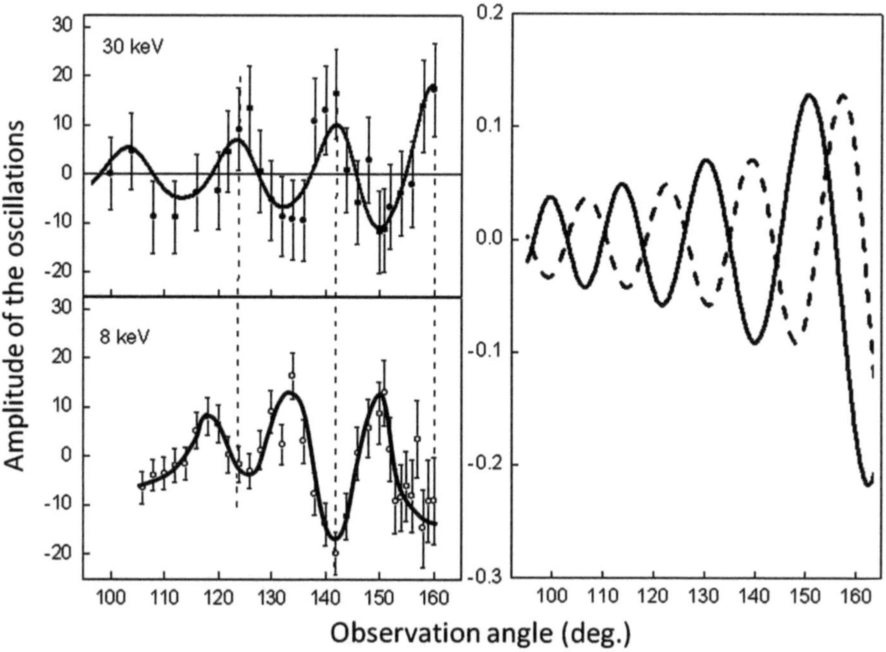

Fig. 4.15 Angular distribution of emitted electrons at 8 and 30 keV. Experiment, *left-hand side*; calculation using path interference model, *right-hand side*. 8 keV, *full curve*; 30 keV, *dashed curve*

periodic variation of the intensity or of the FWHM. A combination of both variations can also be considered.

To answer this fundamental question, the analysis of the Auger spectra was performed in greater detail [8]. First, at each angle, the Auger structures were fitted with appropriate functions. Since the electrons are emitted in the field of the protons, the electron profile, which can be assimilated as a Gaussian curve centered at E_{max} in the absence of protons, is distorted (Barker and Berry effect, cf. Appendix 4). To take into account this distortion, a function of the type $a\ \exp(-b\arctan(c/(E-E_{\max})))$, where a, b and c are free parameters, was multiplied by the Gaussian function (Fig. 4.17). In fact, it could be shown that the distortion mainly acts on the $2s^2\ {}^1S$ configuration.

Then, for the $2s^2\ {}^1S$ configuration, the maximum of intensity I_{max} and the FWHM were extracted as a function of the detection angle (Fig. 4.17). It can be seen in Fig. 4.18 that both the width and the maximum intensity of the $2s^2\ {}^1S$ line oscillate with about the same period, which coincides with the one obtained for the total yield in [3]. There are two other immediately striking results. The first is that the relative amplitude of the linewidth oscillations is comparable to that of the maximum intensity. The second, which is entirely unexpected, seems to indicate that the width and the maximum intensity oscillate in counterphase.

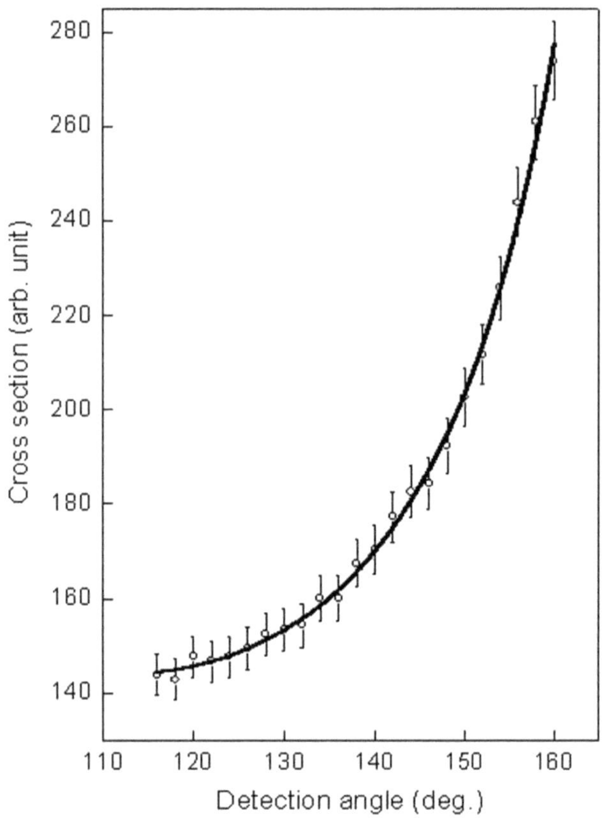

Fig. 4.16 Experimental cross section (*open circles*) for electron emission in 105 keV N^{6+} + H_2 collisions as a function of the detection angle. The *full curve* is the result of a fitting procedure using a 3rd order polynomial function

To explain the observed oscillations and the phase shift, a simple model was developed [8]. The total intensity $I(E, \theta_d)$ is the sum of two contributions, describing the direct ionization and the autoionization, weighted by an interference term

$$I(E, \theta_d) = a(\theta_d) + b(\theta_d)\, j_o(\delta) \tag{4.5}$$

where $j_o(\delta) = \sin\delta/\delta$ is a Bessel function of zeroth order. The width of the profile was determined by the addition of the partial widths $\Delta E_+(\theta_d)$ and $\Delta E_-(\theta_d)$ on each side of the profile maximum E_{max}, using the relation

$$I(E_{\max} \pm \Delta E_\pm, \theta_d) = \frac{1}{2} I(E_{\max}, \theta_d) \tag{4.6}$$

At the first order of the Taylor expansion, the expression of the width is

$$\Delta E(\theta_d) = \Delta E_o(\theta_d) - a(\theta_d) j_o(\delta) + b(\theta_d) j_1(\delta) \tag{4.7}$$

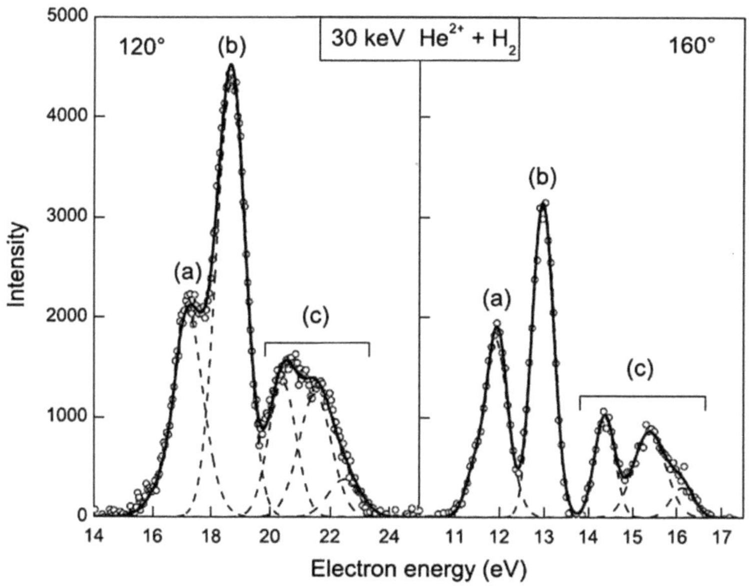

Fig. 4.17 Autoionization spectra in 30 keV He^{2+} + H_2 collisions, for detection angles of 120° and 160°. Peaks a, b and c, see text (Sect. 4.2.3). The *dashed curves* fit the different structures. The *full curve* is the sum of the individual fitted structures

where $j_1(\delta) = \sin\delta/\delta^2 - \cos\delta/\delta$ is the first order Bessel function and $\Delta E_o(\theta_d)$ is the non-oscillating contribution of the width. The functions $a(\theta_d)$ and $b(\theta_d)$ have to be positive to ensure that I_{max} and the FWHM oscillate in counterphase. If, for example, the function $a(\theta_d)$ is negative, the phase shift is only $\pi/4$. The experimental FWHM was therefore fitted using relation (4.7) (full curve at the top of Fig. 4.18). The fitted result agrees with the experiment, since $a(\theta_d)$ and $b(\theta_d)$ were found to be positive. Finally, it is shown that the large amplitude of the width oscillations is mainly due to the contribution of direct ionization.

4.4 Analogy with the Photon Interference Experiment

4.4.1 Relevance of the Analogy

From the beginning, electron and photon interferences were claimed to be analogous. However, differences have certainly to be revealed and pointed out, since the interactions involved are not the same. For example, if a source emits light with wavelength λ within $\Delta\lambda_o$, the final FWHM distribution is expected to be unchanged, i.e. $\Delta\lambda = \Delta\lambda_o$, in contrast with what happens with electrons. However, no photon interference experiment, to our knowledge, has been performed in order to analyze

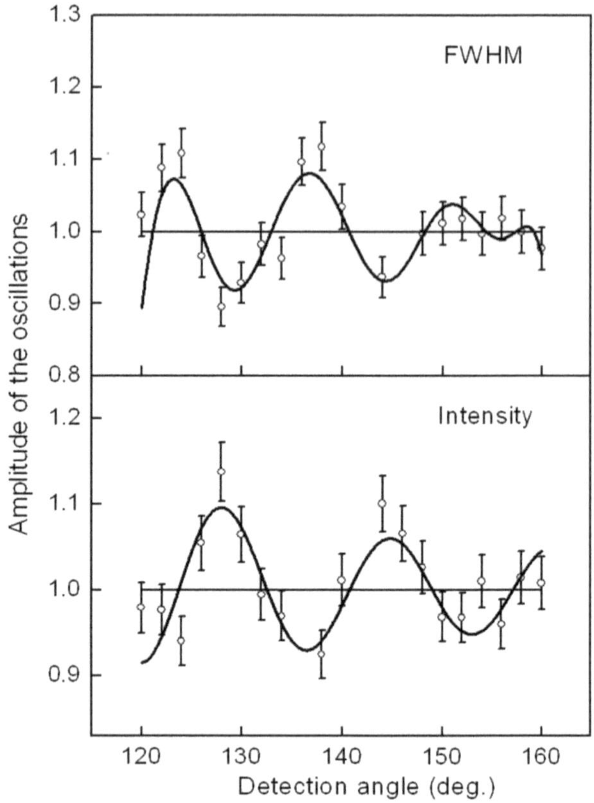

Fig. 4.18 Angular distribution of the oscillation amplitude for the FWHM (*top*) and the intensity at maximum (*bottom*). Experiment, *open circles* ; *full curve*, fit of the experiment using relation (4.6) for the width and a Bessel function of zeroth order for the intensity

the width distribution as a function of the position on a screen. For this reason, the photon interference experiment has very recently been revisited [9].

4.4.2 The Young Double-Slit Experiment with Photons Revisited

4.4.2.1 The Experimental Set-up

The optical setup (Fig. 4.19) is based on the classical Lloyd mirror, instead of the Young double-slit apparatus. This choice was made due to the fact that the luminosity of the Lloyd interference apparatus is much higher. The light source originates from a superluminescent erbium-doped silica fiber. The output profile is a broadband optical emission spectrum (open circles in Fig. 4.18) between 1.5 μm and 1.7 μm (Fig. 4.20). Nevertheless, a peak centered at about 1.53 μm is well separated from the rest of the spectrum. Since $\Delta E/E$ ~ 0.6 μm in the case of the electron interference experiment, an emission peak with a Gaussian profile centered

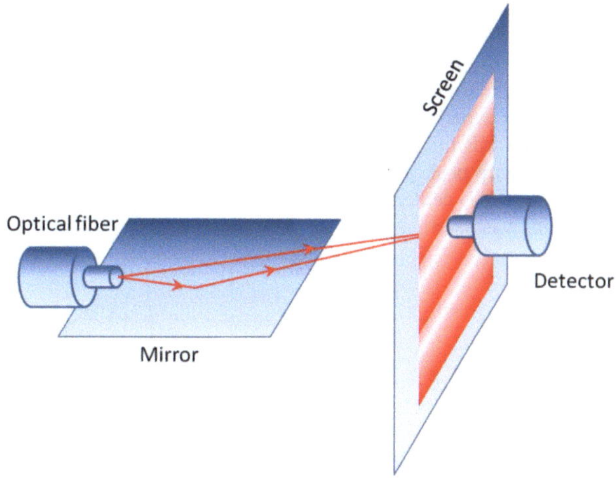

Fig. 4.19 Schematic view of the experimental setup. Light originating from the optical fiber reaches the screen or the detector either directly or after reflection on the mirror, producing interferences

at $\lambda_{max} = 1.532$ μm with a width $\Delta\lambda = 0.93$ μm was selected before the light enters the interferometer (dashed curve in Fig. 4.20).

The combined direct and reflected light give rise to interferences that were detected point by point using a multimode fiber. The collected light was analyzed using an optical spectrum analyzer for simultaneous recording of the spectral and intensity distributions as functions of the detection position.

4.4.2.2 Resulting Oscillations

Figure 4.21 shows a typical wavelength distribution spectrum as a function of the position of the detector. The maxima and minima are well separated, allowing for a detailed analysis of the width distribution.

The upper part of Fig. 4.22 shows a typical experimental interference pattern that consists of very well defined oscillations in the photon intensity distribution. The width oscillates around the average value of 0.93 nm (middle part of Fig. 4.22). Whereas quasi-sinusoidal oscillations were observed in the case of electron interferences, the oscillations were periodic in the case of photon experiments, but not sinusoidal. In addition, in the case of photons, the amplitude of the oscillation is of the order of 1 %, which is much smaller than that determined for electron interferences. Finally, the width and the intensity are found to oscillate in phase. The position λ_{max} also oscillates with the same period (lower part of Fig. 4.22), and relative amplitude of less than 0.01 %. A simple theoretical approach based on classical optics (Appendix 6) makes a good reproduction of the experimental features, as accurately for the width as for λ_{max}.

As a conclusion, it is clear that the present study extends the analogy between photon and electron interference experiments. The observed differences highlighted can be attributed to the different nature of the interactions involved.

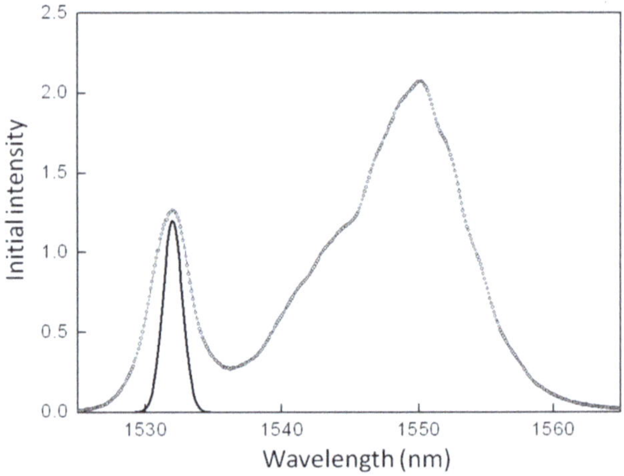

Fig. 4.20 Optical profile as a function of the photon wavelength (*dotted line*). To mimic the electron interference experiment, the distribution was filtered. Only the sharp peak (*full Gaussian curve*) centered at 1.532 μm was used

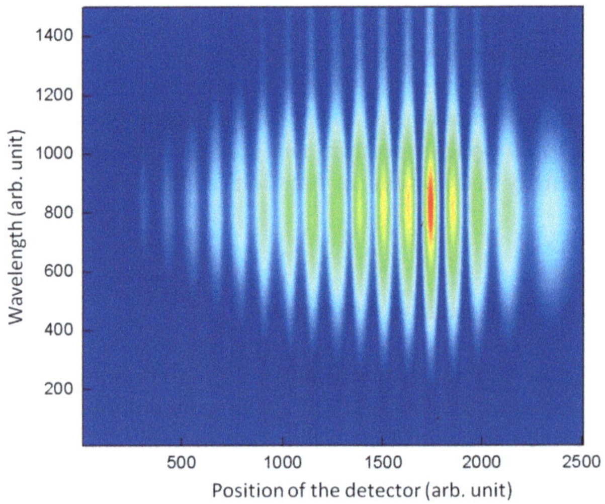

Fig. 4.21 Wavelength distribution after interferences, as a function of the position of the detector

4.4.3 Electron Interferences Revisited

4.4.3.1 E_{max} Angular Distribution

As noted above, even the quantity λ_{max}, defined as the wavelength for which the intensity is maximum, oscillates periodically. Therefore, one wonders if the analogy between electron and photon experiments can be taken any further. In principle, the quantity E_{max}, defined as the electron energy for which the intensity is maximal, is expected to oscillate; however, in practice, identifying this phenomenon is not straightforward. In fact, the detection angle is determined with a relative uncertainty of ~1 %, which is enough to cause the loss of precise information on E_{max}.

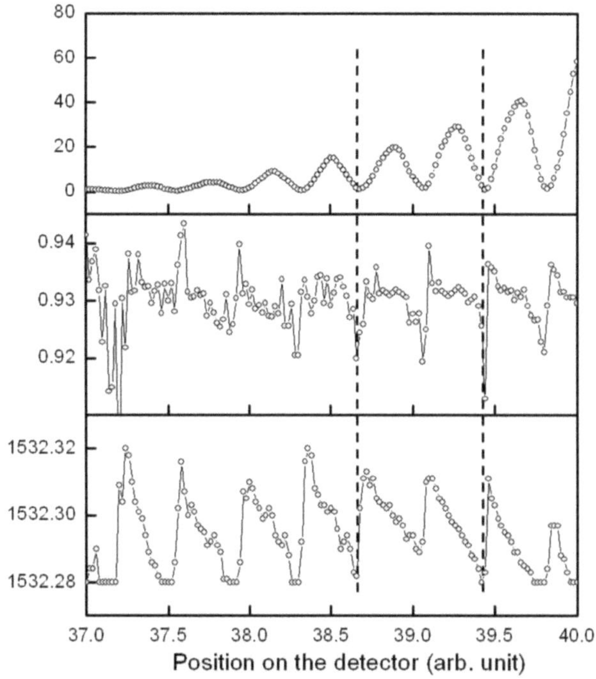

Fig. 4.22 Experimental results: total light intensity (*top*), FWHM (*middle*) and λ_{max} (*bottom*) of the spectral distribution as a function of the relative position of the detector. The relative uncertainties are ~0.01 nm for the FWHM and λ_{max}

This loss can be offset by determining not the energy of the $2s^2\ ^1S$ peak, but the difference between the $2s^2\ ^1S$ and the $2p^2\ ^1D$ peaks (Appendix 7).

4.4.3.2 Energy Distribution at a Fixed Angle

As mentioned in Chap. 3, the energy distribution at a fixed angle is affected by interferences between the different Auger structures. The consequence is that the peaks cannot be fitted using Gaussian curves distorted by a function which describes Post-Collion Effects (PCI) effects. Due to interferences between DI and AI, interferences between Auger lines, and Young-type interferences, the energy profile can be greatly modified. When looking in more detail at a typical Auger spectrum at 160° for example (Fig. 4.23), at which the resolution is the best, oscillations with small amplitude are observed on the left side of the $2s^2\ ^1S$ peak. The positions of the maxima, indicated by arrows on Fig. 4.23, are separated by ~0.3 eV. The pseudo-period of the oscillations can be determined by applying Fast Fourier Transform (FFT) on the energy spectrum (Appendix 8).

Figure 4.24 shows the result of the FFT on the experiment (open circles) at observation angles of 160° and 120°, and on the fitting result (dashed curves), which does not take into account the present oscillations. Three maxima, labeled a, b and c are observed at both angles.

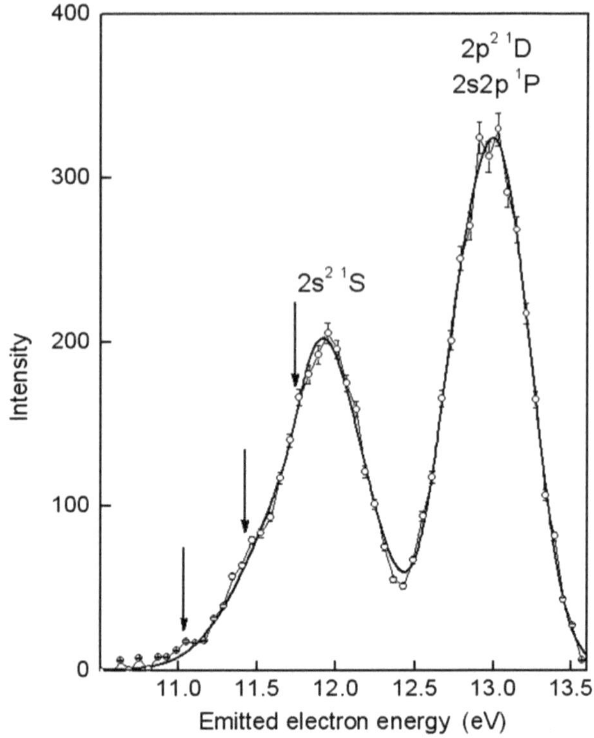

Fig. 4.23 High resolution energy spectrum at an observation angle of 160°. Oscillations are clearly seen (*arrows* on the figure indicate the positions of the maxima)

Structure a is due to the fact that two peaks are involved in the transformation. Indeed, this structure is also present in the calculation. Structure c is caused by the background, which gives rise to pseudo oscillations. Finally, structure b is the signature of physical oscillations, and is due to interferences between the structures (Appendix 8).

4.5 Summary

Young-type interferences in atomic collisions have been studied over the last decade. Scattering of single electrons on two atomic centers induces interferences that manifest themselves in the production of maxima and minima in the angular distribution of electrons, similarly to what happens when photons cross two holes. A detailed analysis of the electron energy profiles and photon wavelength distributions, confirms that it is valid to make an analogy between photon interference and electron interference experiments. The FWHM, intensity, and maximum of energy (or wavelength) for electrons (or photons) were found to oscillate, with the same period as that of the total intensity. However, differences were revealed.

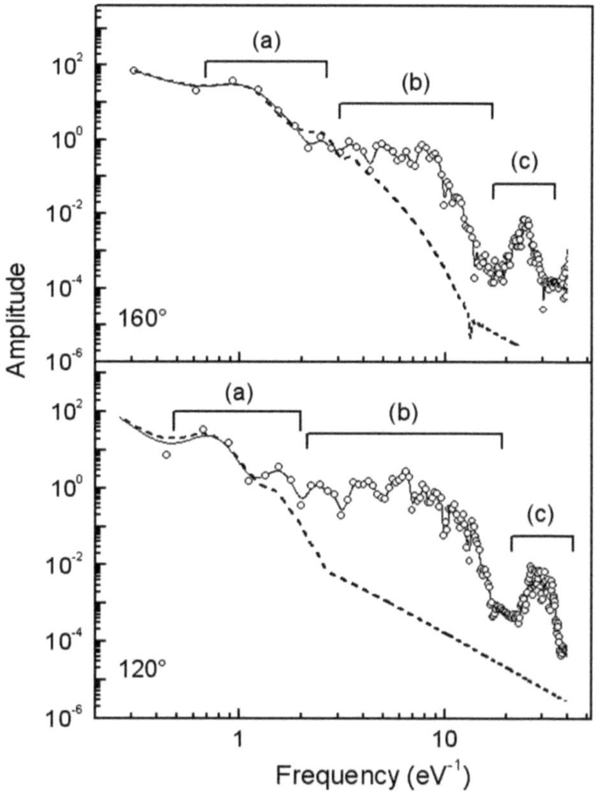

Fig. 4.24 FFT of the experimental spectra (*open circles*) and spectrum fit (*dashed curves*) at observation angles of 160° and 120°. The structure b is due to oscillations in the energy distribution caused by Auger peak interferences

For example, in the case of photon interferences, the angular variation of FWHM and maximum of wavelength are periodic, but not sinusoidal. In addition, while no oscillation could be observed in the wavelength distribution, oscillation in the Auger energy distribution, due to overlapping between the different structures, could be clearly seen.

A question remains concerning the theory underlying electron interferences. In photon experiments, the theory, based on a wave description of photons, reproduces the experimental results well. In electron interferences, the problem is more complex. During the He^{2+} + H_2 collision, five bodies are involved, so the complete theoretical formulation of the problem is not possible. The simple path interference model, described in Sect. 4.2.4, is able to reproduce the main features of the interference phenomenon. Nevertheless, this model is limited to the determination of the oscillation period and shape. It does not give the shape of the Auger profile after the collision. A more complete theoretical analysis is therefore needed. This is the objective of the following chapter.

References

1. Barrachina, R.O., Žitnik, M.: Young's interference effects in the autoionization of atoms colliding with molecules. J. Phys. B **37**, 3847 (2004)
2. Chesnel, J.-Y., et al.: Experimental evidence for Young's interference effects in autoionization following 30 keV He^{2+} + H_2 collision. J. Phys. Conf. Series **58**, 185 (2007)
3. Chesnel, J.-Y., et al.: Young-type experiment using a single-electron source and independent atomic-size two-center interferometer. Phys. Rev. Lett. **98**, 100403 (2007)
4. Frémont, F., et al.: A Young-type experiment using a single-electron source and an independent atomic-size two-center interferometer: the realization of a thought experiment. J. Phys. Conf. Series **88**, 012020 (2007)
5. Debye, P.: Zerstreuung von Röntgenstrahlen. Ann. Phys. (Leipzig) **46**(809) (1915)
6. Ehrenfest, P.: Amsterdam Acad. **23**, 1132 (1915)
7. Frémont, F., et al.: Interférences de type Young à un seul electron. C. R. Physique. **9**, 469 (2008)
8. Barrachina, R.O., et al.: Linewidth oscillations in a nanometer-size double-slit interference experiment with single electrons. Phys. Rev. A **81**, 060702(R) (2010)
9. Vabre, M., et al.: Periodic variations in the wavelength distributions following photon interferences: analogy with electron interferences, ISRN Spectroscopy Volume 2012, Article ID 174952, 4 pages

Chapter 5
A Theoretical Description of Young-Type Interferences Following Auger Electron Emission

5.1 Quantum Mechanical Description

5.1.1 Collision Diagrams

The theoretical Auger formalism has already been done for ion colliding on atoms, so that the goal of the present study was to extend this to ion-molecule collisions. These types of collisions will, therefore, be theoretically compared. To simplify, the collision systems He^{2+} + He and He^{2+} + H_2 will be taken as examples (Figs. 5.1 and 5.2).

Contrary to the common view, where the He^{2+} ion moves with velocity $\overrightarrow{v_P}$ and the target is at rest, the projectile is supposedly at rest, with two electrons in a $2\ell 2\ell'$ doubly excited state, and the target, He^{2+} or H_2^{2+}, which has lost both electrons, moves with velocity $-\overrightarrow{v_P}$. Therefore, the electron is emitted with a momentum $\overrightarrow{k_T}$ (or $\overrightarrow{k_P}$) in the reference frame of the target, i.e. He (or of the projectile, i.e. He^{2+} or H_2^{2+}), as seen in Figs. 5.1 and 5.2. In both cases, it is assumed that the target has captured the two electrons from He or H_2 at time $t = t_o$, where $t_o = OP_o/v_N$. In addition, the projectile is assumed to have, during the whole process, a straight-line trajectory, so that

$$\overrightarrow{r_N} = \overrightarrow{b} - \overrightarrow{v_P}\,(t + t_o) \tag{5.1}$$

where b is the impact parameter.

In the case of H_2, both protons dissociate and have a velocity v_H at infinite distance. The distance between the H_2 center of mass P and each proton is derived from the approximated formula

$$r_H \approx r_{H_o} + v_H t \tag{5.2}$$

F. Frémont, *Young-Type Interferences with Electrons*, Springer Series on Atomic, Optical, and Plasma Physics 77, DOI 10.1007/978-3-642-38479-0_5,

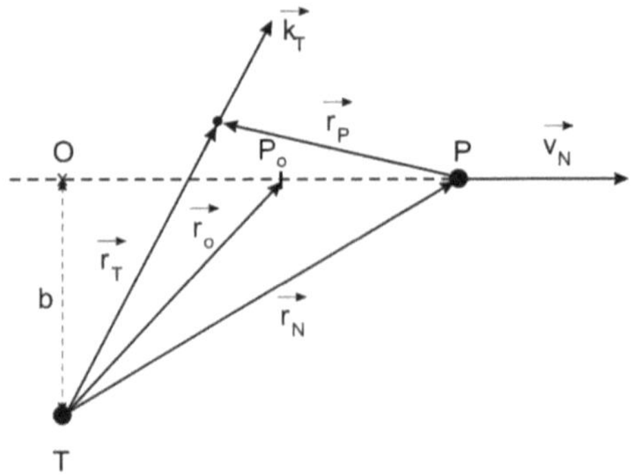

Fig. 5.1 Schematic diagram of the collision between an atomic ion *P* and an atom *T*. The capture occurs at P_o

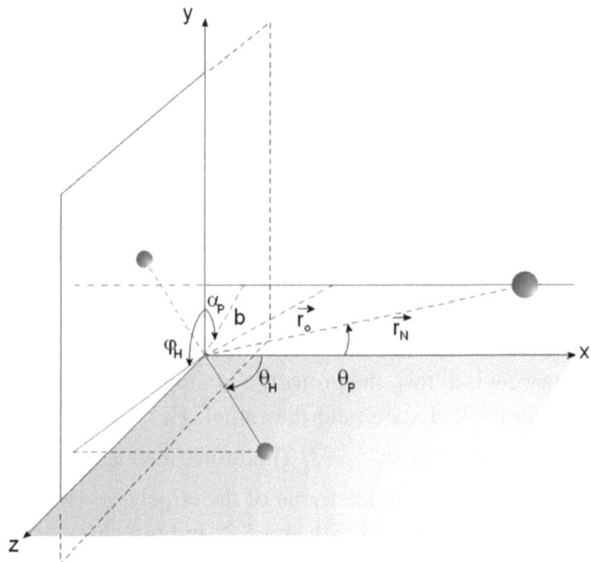

Fig. 5.2 Schematic diagram of the collision between an molecular ion characterized by P_1 and P_2 and an atom *T*

where $2\ r_{H_o} = 1.42$ a.u. is the distance between the protons at equilibrium. The position in space of the molecule is characterized by its angles θ_H and φ_H.

5.1.2 Basic Quantum Mechanics Formalism

The autoionization process consists of two-steps. First, the two electrons from *P* are captured by *T*. After the capture, and independently from this primary process, the autoionization of *T* occurs. The first model proposed to describe this two-step process was presented by Barker and Berry [1] in 1966, based on phenomenological considerations. The first quantum-mechanical model was developed eleven years later [2] and assumed that only the energy of the electron, but not its trajectory, is changed due to the vicinity of the projectile. In 1986, a semi-classical model [3] was introduced, taking into account the deviation of the electron, as well as the interferences between the structures. Nevertheless, despite improvement offered by this model, the angular distribution of emitted electrons was found to be incorrect for small angles. In 1989, a model based on Continuum Distorted Wave (CDW) approximation was introduced [4], and was found to incorporate the previous models as particular cases in different regions of the electron and projectile velocity space, as well as electron angles.

Assuming that the target is in an autoionizing state $|\psi_i\rangle$ at $t = t_o$, the wavefunction at $t > t_o$ is given by

$$|\psi(t)\rangle = a(t)\exp\left(-i\int_0^t E(t')\,dt'\right)|\psi_i\rangle + \int b(t)\exp(-k_T^2/2)\,|\psi_{k_T}\rangle\,dk_T \tag{5.3}$$

Replacing $|\psi(t)\rangle$ in the time-dependent Schrödinger equation, two coupled equations are obtained

$$\frac{da(t)}{dt} = -i\int_0^\infty \langle\psi_i|H_i|\psi_{k_T}\rangle \exp\left(-i\int_0^t (k_T^2/2 - E(t'))\,dt'\right) b(t)\,dk_T \tag{5.4a}$$

$$\frac{db(t)}{dt} = -i\,\langle\psi_{k_T}|H_i|\psi_i\rangle\,\exp\left(i\int_0^t (k_T^2/2 - E(t'))\,dt'\right) a(t) \tag{5.4b}$$

where H_i is the interaction Hamiltonian.

The probability amplitudes $a(t)$ (or $b(t)$) are determined by integration of expressions (5.4a) (or (5.4b)), and replacing the result in Eq. 5.4b (or Eq. 5.4a). Using a peaking approximation, both previous expressions can be simplified; for $a(t)$ and $b(t)$, this leads to

$$a(t) = \exp\left(-\int_0^t \Gamma(t')/2\,dt'\right) \tag{5.5a}$$

$$b(t) = -i\int_0^\infty \langle\psi_{k_T}|H_i|\psi_i\rangle\,\exp\left(i\int_0^t (k_T^2/2 - E(t') + i\Gamma(t')/2)\,dt'\right) dt \tag{5.5b}$$

In the whole of this section, $\Gamma(t')$ is supposedly independent of time, so that it is a constant that characterizes the autoionizing state. In fact, in the most general case, $\Gamma(t')$ strongly depends on time (see for example [5]).

Since the electron is disturbed by the presence of the projectile, its energy E_o changes and becomes

$$E(t) = E_o - \frac{2}{r_N(t)} \tag{5.6}$$

Thus, the amplitude of transition probability is

$$b(t) = -i \int_0^\infty \langle \psi_{k_T} | H_i | \psi_i \rangle \, e^{i \int_0^t -\frac{2}{r_N(t')} dt'} \, e^{i\left(k_T^2/2 - E_o + i\Gamma/2\right)t} \, dt \tag{5.7}$$

The incorporation of the interaction between the emitted electron and the projectile is made by distorting the ingoing final continuum state [4], in order to separate the influences of the projectile and target on the electron.

$$\left| \psi_{k_T}(t) \right\rangle \approx D_{-k_P}\left(\overrightarrow{r_P}\right) \left| \psi^*_{k_T}\left(\overrightarrow{r_T}\right) \right\rangle \tag{5.8}$$

In this expression, $\left|\psi_{k_T}\left(\overrightarrow{r_T}\right)\right\rangle$ is the Coulomb continuum wavefunction of the electron, moving with the asymptotic momentum $\overrightarrow{k_T}$, in the field of the residual target ion only, and $D_{-k_P}\left(\overrightarrow{r_P}\right)$ is a distortion factor that takes into account the proximity of the projectile; therefore

$$\langle \psi_{k_T} | H_i | \psi_i \rangle \approx D_{k_P}\left(\overrightarrow{r_P}\right) \left\langle \psi^*_{k_T}\left(\overrightarrow{r_T}\right) \right| H_i | \psi_i \rangle.$$

In addition, it is assumed that the matrix element $\left\langle \psi^*_{k_T}\left(\overrightarrow{r_T}\right) \right| H_i | \psi_i \rangle$ is not equal to zero only when the electron is close to the target, i.e. $\overrightarrow{r_P} \approx -\overrightarrow{r_N}$. Therefore, the amplitude of probability is written

$$b(t) = -i \left\langle \psi^*_{k_T} | H_i | \psi_i \right\rangle \int_0^\infty D_{k_P}\left(\overrightarrow{r_N}\right) e^{i \int_0^t -\frac{2}{r_N(t')} dt'} \, e^{i\left(k_T^2/2 - E_o + i\Gamma/2\right)t} \, dt \tag{5.9}$$

5.1.3 *Application to Ion-Atom Collisions*

5.1.3.1 Autoionization in the Absence of Projectile

The integrand in expression (5.9) consists of three terms. The last of these terms, namely $e^{i\left(k_T^2/2 - E_o + i\Gamma/2\right)t}$, describes the autoionization of the target in the absence of projectile. This is achieved by making $D_{k_P}\left(\overrightarrow{r_N}\right) = 1$ and $e^{i \int_0^t -\frac{2}{r_N(t')} dt'} = 1$. Expression (5.9) can be simplified and it follows that [6,7]

$$b(t) \propto \frac{e^{i\left(k_T^2/2 - E_o + i\Gamma/2\right)t}}{i(k_T^2/2 - E_o + i\Gamma/2)} \tag{5.10}$$

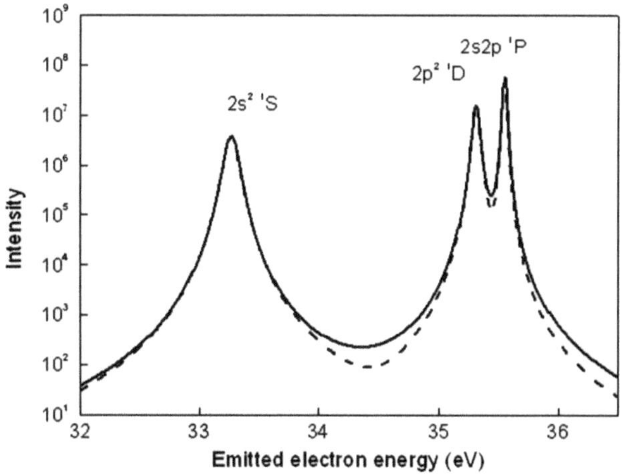

Fig. 5.3 Calculated spectra for autoionizing structures associated with $2s^2\ ^1S$, $2p^2\ ^1D$ and $2s2p\ ^1P$ configurations. *Dashed curve*: the total intensity is the sum of individual intensities; *full curve*: the interferences between the peaks are taken into account

The shape of the electron profile is a Lorentzian whose FWHM is Γ. Figure 5.3 shows the result of the calculation for the three He configurations $2s^2\ ^1S$, $2p^2\ ^1D$ and $2s2p\ ^1P$. The dashed line represents the sum of the intensities, whereas the full curve is the result, taking into account the interferences between the three peaks. It is seen that the differences between both results are small compared with the intensity. In this case, interferences can be neglected.

In many experiments, the electron is detected in coincidence with the recoil target ion and the projectile. As a consequence, the capture parameters, for example the recoil angle (i.e., the angle between the recoiling target ion trajectory and the incident beam trajectory), scattering angle (i.e., the angle between the scattered projectile trajectory and the incident beam trajectory), and Q-value (i.e., the projectile energy difference before and after the collision), are directly connected with the parallel and perpendicular components $k_{T//}$ and $k_{T\perp}$ of the electron momentum, with respect to the incident beam direction. In momentum space, the intensity for the three peaks consists of rings whose maximum of intensity is centered on 1.56–1.62 a.u., as shown in Fig. 5.4. No additional structure is seen in this figure, since the interferences between direct ionization and autoionization, or the deflection of the electron due to the presence of the two protons, are not taken into account.

5.1.3.2 Distortion of the Auger Profile in the Field of the Projectile

The first two terms in the integrand of relation (5.9) have two distinct consequences. The term $e^{i\int_0^t -\frac{2}{r_N(t')}dt'}$ describes the autoionization in the field of the projectile. Its introduction distorts the electron energy profile, while the term $D_{k_P}\left(\overrightarrow{r_N}\right)$ distorts the electron wavefunction. If we take into account only the second term, the amplitude becomes

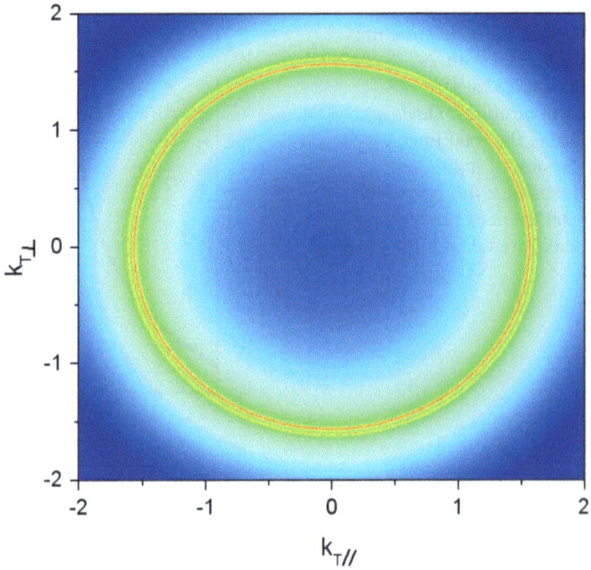

Fig. 5.4 Calculated intensities in momentum space for autoionizing structures associated with $2s^2\ ^1S$, $2p^2\ ^1D$ and $2s2p\ ^1P$ configurations

$$b(t) = -i\ \left\langle \psi^*_{k_T} \middle| H_i \middle| \psi_i \right\rangle \int_0^\infty e^{i\int_0^t -\frac{2}{r_N(t')}dt'}\ e^{i\left(k_T^2/2 - E_o + i\Gamma/2\right)t}\, dt \tag{5.11}$$

Impact parameter distributions for the three configurations were determined theoretically [8] at projectile energies larger than 100 keV. It has been shown (Fig. 5.5) that the average impact parameter for $2\ell 2\ell'$ configurations is ~1 a.u. According to Fig. 5.5, the crossing radius at which the capture occurs is, close to 2 a.u., so that t_o is of the order of 4 a.u.

With these parameters, calculation was performed at a detection angle of 20° (Fig. 5.6).

The profile of each structure is distorted, and a tail is observed on the left side of the peaks, due to the presence of the projectile. Differences are also observed according to whether interferences between the peaks are taken into account (full curve) or not (dashed curve). In particular, the tail for the $2p^2\ ^1D$ configuration is seen to be less pronounced, while changes in $2s^2\ ^1S$ and $2s2p\ ^1P$ configurations are minor.

The influence of time t_o is represented in Fig. 5.7, for an impact parameter of 1 a.u. When t_o increases, the $2s^2\ ^1S$ peak shifts to higher energies, and the width decreases. On the contrary, the $2p^2\ ^1D$ and $2s2p\ ^1P$ peaks are not sensitive to time variation in the range from 0.1 a.u. to 10 a.u., and the profile shifts to lower energies at $t_o = 100$ a.u.

Figure 5.8 shows the evolution of the Auger profiles with the impact parameter at $t_o = 4$ a.u. Except for the case $b = 10$ a.u., the profile is nearly independent on b. Finally, since capture occurs at small values of b and t_o, the calculations can be performed assuming $b = 0$ and $t_o = 0$.

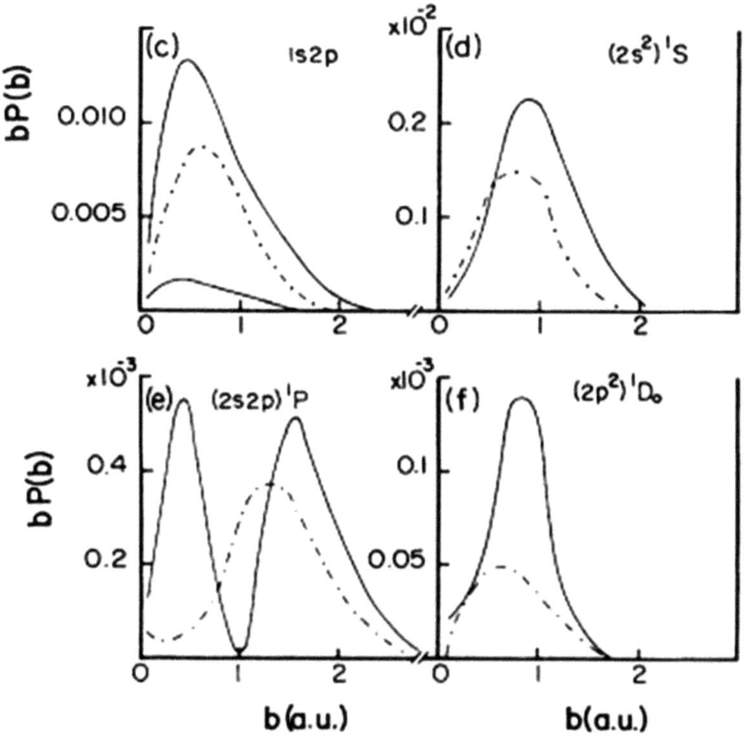

Fig. 5.5 Calculated impact parameter distributions following double electron capture in 100 keV He^{2+} + He collisions, for singly and doubly excited states. *Dashed and full curves* are the results using different models [8]

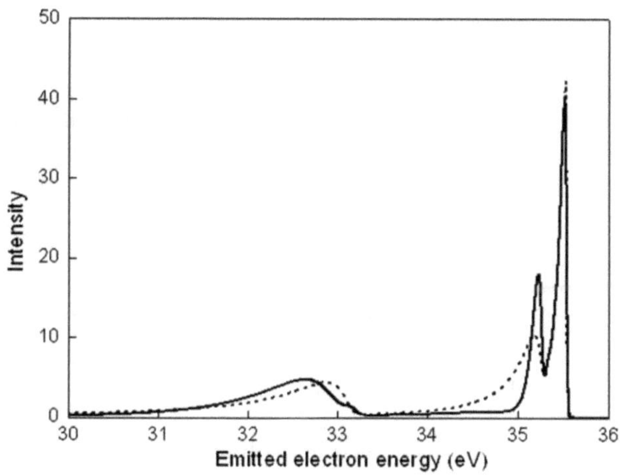

Fig. 5.6 Calculated electron energy profile for the $2s^2$ ^1S, $2p^2$ ^1D and 2s2p ^1P configurations. The quantities b and t_o are equal to 1 and 4 a.u., respectively. *Dashed curve*: Sum of the three contributions; *full curve*: Interferences between the three peaks are taken into account

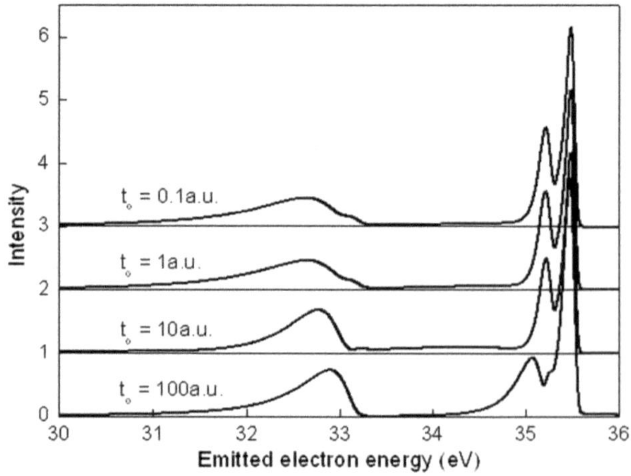

Fig. 5.7 Influence of time t_o on the electron profile, for an impact parameter of 1 a.u

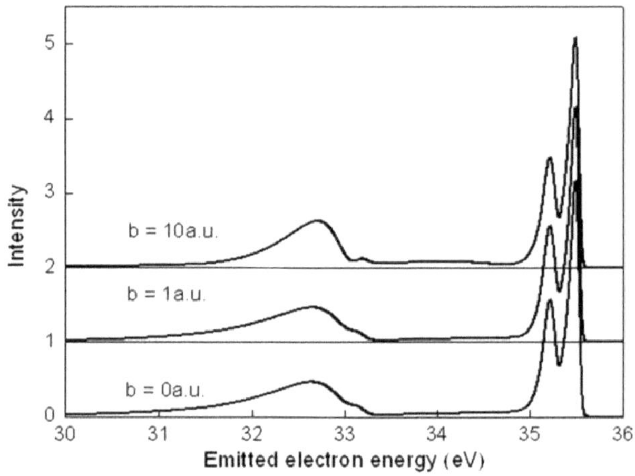

Fig. 5.8 Influence of the impact parameter on the electron profile, for time $t_o = 4$ a.u

The noteworthy feature of these last two models is that there is no dependence on angle, since only the electron energy is changed. To observe an angular dependence, the distortion factor needs to be incorporated.

5.1.3.3 Distortion of the Wave Function

Eikonal Approximation

Several expressions for $D_{k_P}\left(\overrightarrow{r_N}\right)$ appear in the literature. The simplest one is based on the eikonal approximation, leading to [7]

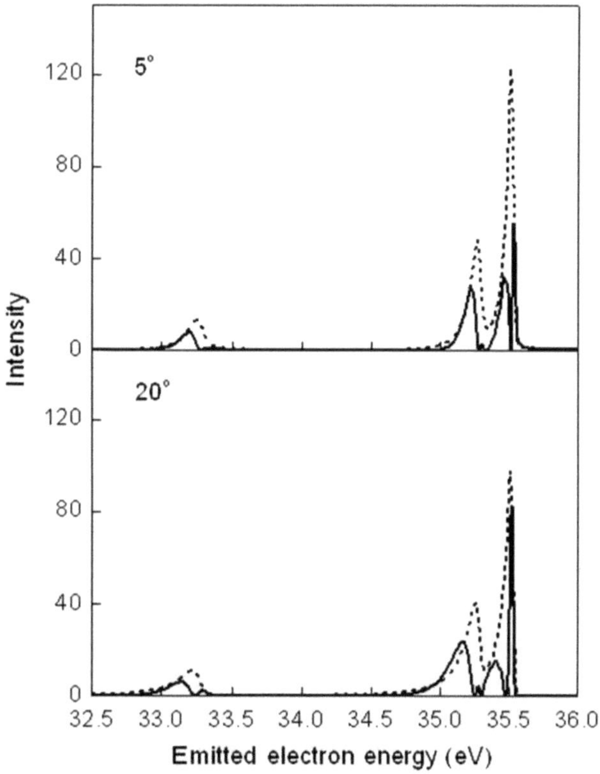

Fig. 5.9 Calculated electron energy profile for the three configurations using eikonal approximation. respectively. *Dashed curve*, incoherent sum; *full curve*, coherent sum

$$D_{k_P}\left(\overrightarrow{r_N}\right) = \chi^{-i\nu_P} \tag{5.12}$$

where $\chi = k_P\, r_N - \overrightarrow{k_P}.\overrightarrow{r_N}$, $\nu_P = 2/k_P$ and $\overrightarrow{k_P}$ is the electron momentum in the projectile frame. Since $\overrightarrow{k_P} = \overrightarrow{k_T} - \overrightarrow{v_P}$, the modulus of $\overrightarrow{k_P}$ is given by

$$k_P = \left(k_T^2 + v_P^2 - 2k_T v_P \cos\theta_d\right)^{1/2} \tag{5.13}$$

The quantity χ (relation (5.12)) is the phase difference obtained in the path-interference model (cf. Appendix 5). Thus, the introduction of $D_{k_P}\left(\overrightarrow{r_N}\right)$ will, in principle, give rise to interferences between the direct path and the electron path, incorporating elastic deflection by the projectile.

Figure 5.9 shows two typical spectra calculated at 5° and 20°. Again, the dashed curves (or full curves) are the result of adding the three contributions incoherently (or coherently). In contrast with the two previous models, the angle dependence can now be observed, in the shape of the structures as well as in the intensity. Compared with the previous models, additional structures are seen for the $2p^2\ ^1D$ and $2s2p\ ^1P$ configurations, due to interferences between the peaks.

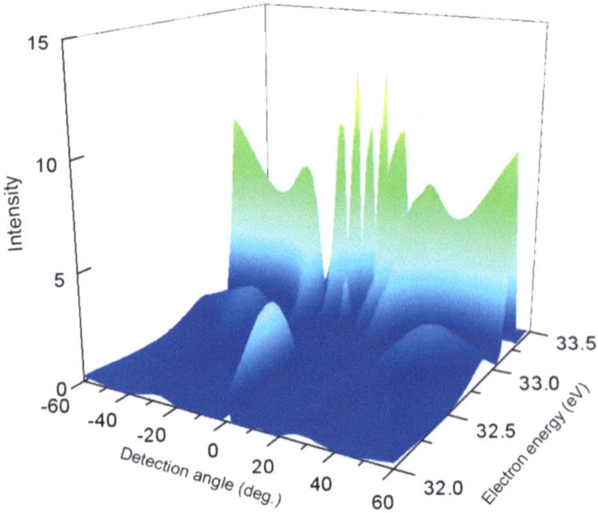

Fig. 5.10 3D plot of the $2s^2$ ^1S in angle and energy, in the range from −60° to 60°

To look at the angular dependence in more detail, a 3D plot of the calculated spectrum associated with $2s^2\ ^1S$ was made as a function of the angle and the emitted electron energy, shown in Fig. 5.10. Strong maxima are observed around 0°. The cross section then decreases at ~10°, and other maxima appear at larger angles.

In momentum space, a typical contour plot is shown in Fig. 5.11, for the three peaks. In addition to the main ring (red full curve), centered on (0,0), structures due to interferences are observed in a portion of ring (dashed curve) centered at (0.63,0).

If interferences between the peaks are not taken into account, and for $b = 0$ and $t_o = 0$, the 3D figure can be simplified (Fig. 5.12)

For this set of b and t_o values, Eq. 5.11 can be simplified to

$$b(t) \approx \frac{\Gamma(1 + i(\nu_N - \nu_P))}{(1 - i\varepsilon)^{i(\nu_N - \nu_P)}} \frac{1}{1 - i\varepsilon} \tag{5.14}$$

with $\varepsilon = 2(E - E_o)/\Gamma$.

A maximum is obtained at 0°, then the intensity decreases with increasing angle. The angular dependence is due to the term $\nu_N - \nu_P$ in expression (5.14). It can be noted that the eikonal approximation does not take into account the possibility for the electron to change its trajectory.

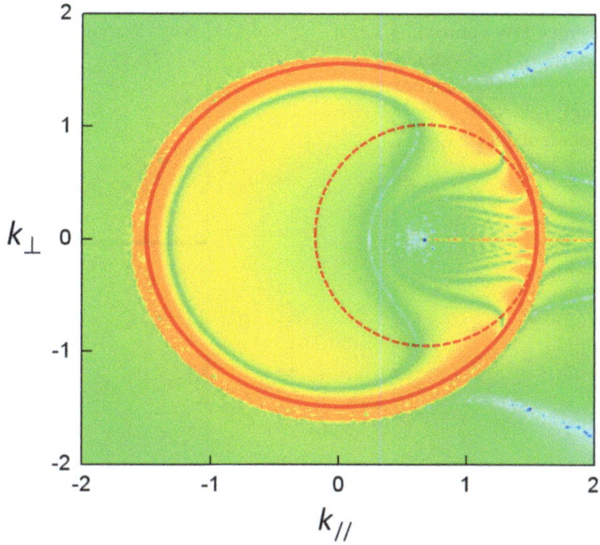

Fig. 5.11 3D contour plot of Auger intensity in the eikonal approximation

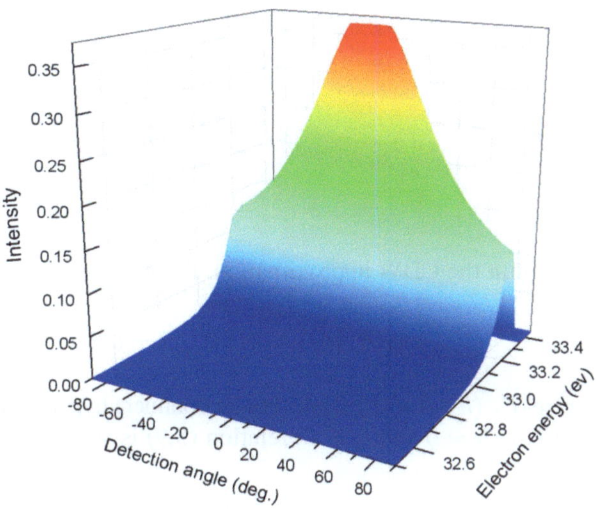

Fig. 5.12 3D plot of the 2s2 1S in angle and energy, for b = 0 and to = 0

Continuum-Distorted Wave Approximation

When the electron moves in the direction of the projectile with comparable velocity, the Coulomb interaction between the electron and the projectile can no longer be considered a perturbation. The continuum-distorted wave (CDW) approximation, which takes into account the electron emission in both projectile and target fields, is therefore used to take into account the strong effect at forward angles. This approximation induces a dependence on the emission angle that was missing from previous theories [4].

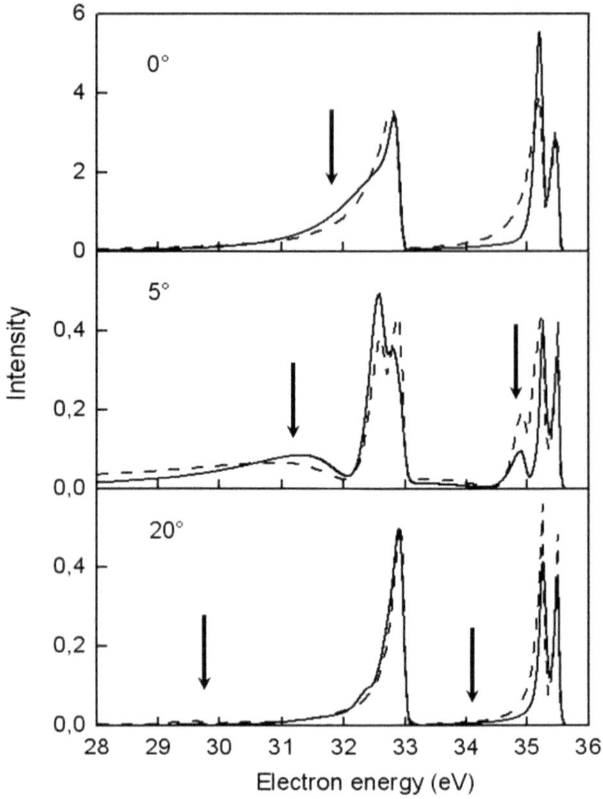

Fig. 5.13 Calculated spectra using CDW approximation at angles of 0°, 5° and 20°. *Dashed curves*: incoherent sum; *full curves*: coherent sum

Within the CDW approximation, the distortion factor can be written

$$D_{k_P}\left(\overrightarrow{r_N}\right) = e^{\pi \nu_P/2}\Gamma(1 - i\nu_P)\ {}_1F_1\ (i\nu_P, 1, i\chi) \tag{5.15}$$

where ${}_1F_1\ (i\nu_P, 1, i\chi)$ is the first order confluent hypergeometric function (Appendix 10).

Using $b = 0$ and $t_o = 0$, relation (5.9) is simplified and becomes

$$A_{CDW} = e^{\pi \nu_P/2}\Gamma(1 - i\nu_P)\frac{\Gamma(1 + i\nu_N)}{(1 - i\varepsilon)^{1+i\nu_N}}\ {}_2F_1\left(i\nu_P, 1 + i\nu_N, 1, \frac{i\xi}{1 - i\varepsilon}\right) \tag{5.16}$$

where ${}_2F_1\ (i\nu_P, 1, i\chi)$ is the hypergeometric function, and the quantity ξ is defined by $\xi = \frac{2}{\Gamma}\left(k_P v_N - \overrightarrow{k_P}\,.\,\overrightarrow{v_N}\right)$

Figure 5.13 shows three calculated spectra at detection angles of 0°, 5° and 20°. As above, dashed curves and full curves represent the incoherent and coherent sums, respectively. First, it should be noted that, at small angles, the spectra do not diverge at energies smaller than 30 eV, in contrast to the spectra calculated using

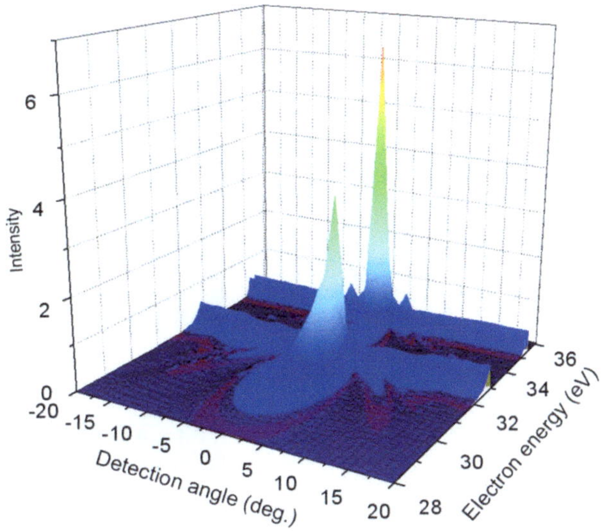

Fig. 5.14 Angular distribution for the coherent sum calculated within the CDW approximation

other approximations (Fig. 5.15). Second, the interferences between the structures do not shift the energies, but increase or decrease the relative intensities. If we focus on the left side of the peaks, some notable features can be seen. Due to electron scattering on the projectile, structures appear whose maxima positions (indicated by arrows) depend on the detection angle. These structures, as mentioned previously, are dominant at forward angles.

Due to the introduction of the hypergeometric function to the CDW approximation, the angular distribution of emitted electrons is peaked at 0°, as seen in Fig. 5.14.

Finally, the total intensity for the $2s^2\ {}^1S$ configuration in momentum space is presented in Fig. 5.15. As in the previous cases, both direct and indirect contributions are observed, as well as the interferences, which are mainly located at forward angles in the target reference frame.

These results are consistent with the experimental data, suggesting that CDW approximation is able to be used in the case of ion-molecule collisions. Nevertheless, for the sake of simplicity, the use of different methods will be discussed as done for ion-atom collisions.

5.1.4 *Application to Ion-Molecule Collisions*

5.1.4.1 Expression for the Transition Amplitude

In ion-atom collisions, the electron energy in the presence of He target writes $E = E_o - 2/r_N$. Replacing the He target by two protons separated by the distance $2r_H$, the electron energy in the field of the protons is

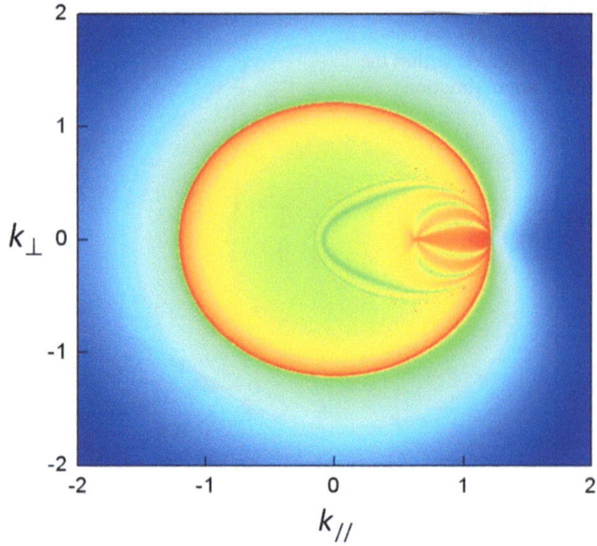

Fig. 5.15 Contour plot of the $2s^2\ ^1S$ contribution in momentum space, calculated using CDW approximation

$$E = E_o - \frac{2}{\left|\overrightarrow{r_N} + \overrightarrow{r_H}\right|} - \frac{2}{\left|\overrightarrow{r_N} - \overrightarrow{r_H}\right|} \tag{5.17}$$

To calculate the autoionization probability, the CDW approximation is used [9]. According to the authors, the autoionization cross section is given by

$$\frac{d\sigma}{dv} \propto \left| \int_0^\infty D(r_N)\, t^{i\nu_N}\, e^{-(1-i\varepsilon)\Gamma t/2} dt \right|^2 \tag{5.18}$$

In this expression, $\varepsilon = \frac{2}{\Gamma}(E - E_o)$ and

$$D(r_N) = 1 + \sum_{n=0}^{1} (D_n - 1)\, e^{i\left(\overrightarrow{k_P} - k_P \overrightarrow{r_N}/r_N\right) . \overrightarrow{r_{Hn}}} \tag{5.19}$$

where r_{H1} and r_{H2} are the distance between each proton and the molecule center of mass.

In CDW approximation, the distortion factor D_n is written

$$D_n = \Gamma(1 + i\nu_N)\, e^{\pi/k_P}\, {}_1F_1\left(i\nu_N, 1 + 2i/k_P, \frac{i\xi}{1 - i\varepsilon} \right) \tag{5.20}$$

where $\xi = \frac{2}{\Gamma}\left(k_P v_P - \overrightarrow{k_P} . \overrightarrow{v_P}\right)$

By incorporating expression (5.19) into (5.18), for the cross section we obtain

$$\frac{d\sigma}{dv} \propto \left| \int_0^\infty \left(1 + 2\cos\left[\left(\overrightarrow{k_P} - k_P\overrightarrow{r_N}/r_N\right).\overrightarrow{r_H}\right](D_n - 1)\right) t^{i\nu_N} e^{-(1-i\varepsilon)\Gamma t/2} dt \right|^2 \tag{5.21}$$

The interferences are given by the term $\cos\left[\left(\overrightarrow{k_P} - k_P\overrightarrow{r_N}/r_N\right).\overrightarrow{r_{Hn}}\right]$, which evolves in time due to the quantity $\overrightarrow{r_H}$.

5.1.4.2 Calculation Results and Comparison with Experiments

As seen previously, the distribution consists, in momentum space, of two circles, one centered at $k = 0$, assuming the projectile at rest, with a radius equal to 1.56 a.u., and a second one centered at v_N, with a radius equal to $1.56 - v_N$. If Young-type interferences exist and are visible, one would expect to see series of maxima and minima on the large and the small rings, as visible in Fig. 5.16.

Using $b = 0$ and $t_o = 0$, the angular distribution in the momentum space could be calculated in the CDW approximation and is presented in Fig. 5.17. Surprisingly, even if the two rings are clearly observed, no oscillation is visible on either the large or the small angle. Instead, interferences manifest themselves in the production of maxima and minima inside the small ring, especially in the forward direction, i.e., for $k_{//} \sim 2$ a.u. and $k_\perp \sim 0$. These oscillations are the signature of interferences between the two rings, but do not provide any evidence for Young-type interferences.

Integrating the calculated spectra over electron energy, the angular distribution was determined and is presented (full red curve) in Fig. 5.18. Also, the calculated distribution is compared to the experimental one (open circles). In contrast with the experiment, the calculation does not show any oscillation, except between 165° and 180°. This oscillation is the signature of the interferences between an electron that reaches the detector directly, and an electron that reaches the detector after it has scattered on one proton.

In conclusion, the present model is not able to reproduce the experimental oscillations. One might ask what influence the distortion factor has on the results. In fact, whatever this factor, one expects no change in the present results. Rather, the problem originates from the integral in time itself. Relation 5.20 shows that time varies from 0 to infinity. This induces distances between the protons that also vary from 1.42 a.u. (equilibrium distance) to infinity. Consequently, the total amplitude consists of an infinite number of individual amplitudes that vary periodically, with a pseudo-period that changes continuously (cf. Appendix 11). The infinite numbers of oscillations lead to a strong decrease in the contrast.

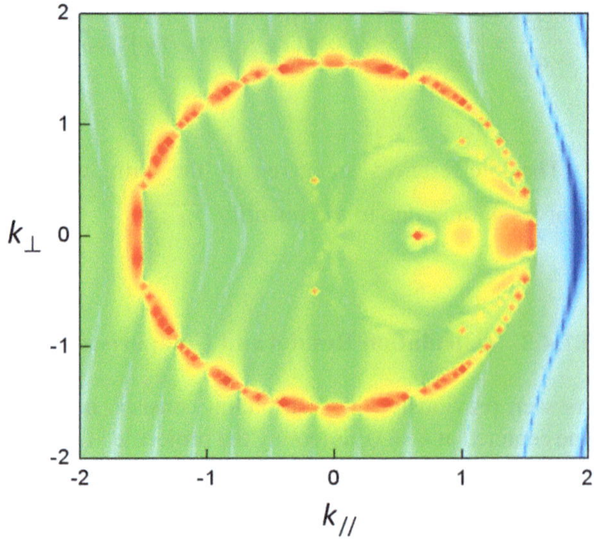

Fig. 5.16 Contour plot of the angular distribution in momentum space expected when Young type interferences are visible. The figure was built using the formula given in [4]

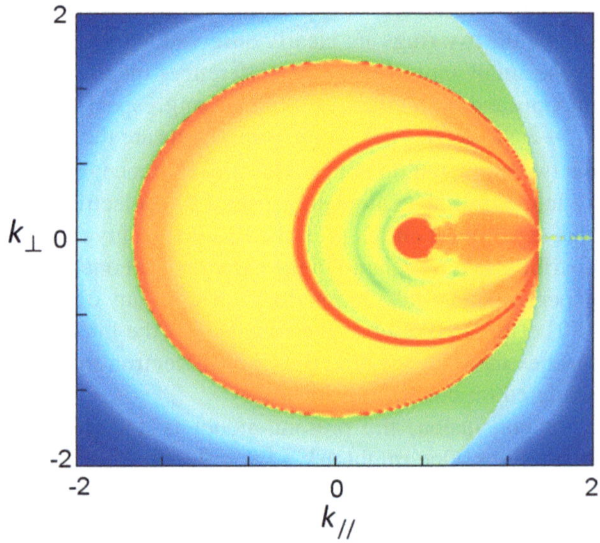

Fig. 5.17 Contour plot of the angular distribution in momentum space calculated using CDW approximation. The two rings are observed; as well as the interferences between direct and indirect path (*green curves inside the small circle*)

5.2 A Semi-classical View of Interferences

5.2.1 Introduction

In the last section we saw the details of how autoionization theory using the Final State Interaction approximation does not reproduce the experimental angular distribution of emitted electrons. The path interference model, based on the difference between the

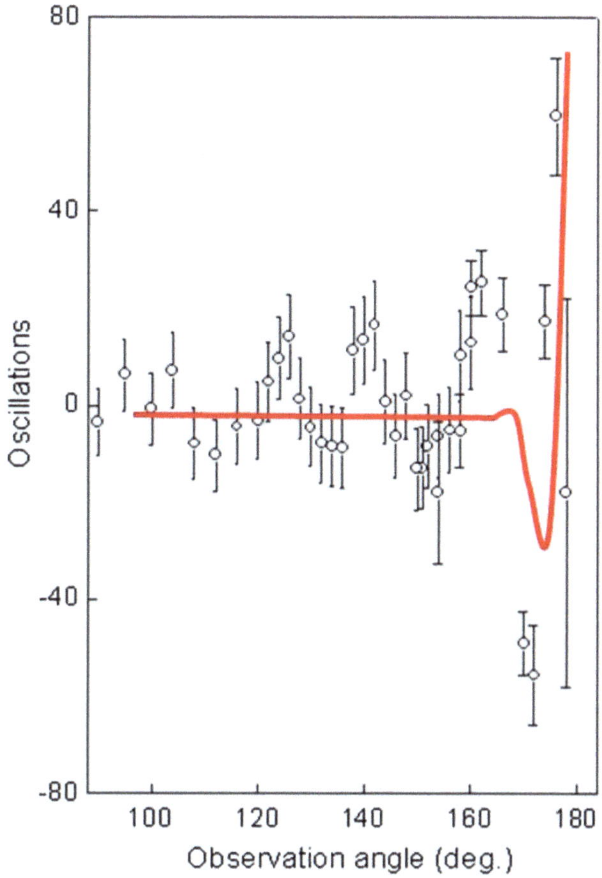

Fig. 5.18 Oscillations obtained experimentally (*open circles*) and theoretically (*red full curve*), using CDW. One oscillation is observed at angles larger than 165°

various paths taken by the electrons, reproduces the experimental angular distribution, but is unable to give the energy distribution of the electrons at a given detection angle. In addition, this model assumes that the electrons are ejected at an average distance determined by the lifetime of the Auger structure. In other words, the exponential variation of electron emission is not taken into account. Because no quantum approach exists at present to treat this complex problem, a semi-classical approach was taken. The idea for the following treatment originated from macroscopic-scale experiments performed a few years ago [10].

In the experiment, a droplet of silicon oil falls on a vertically vibrated bath of the same fluid. Under specific conditions, the droplet becomes self-propelled and moves indefinitely on the liquid surface, with a constant velocity, generating waves. This moving droplet, dressed with the wave-packet it emits, the authors named a "walker".

The walker is thus constituted of a corpuscle and a wave. Neither the wave nor corpuscular aspects can exist separately. If the wave disappears, the walker stops and the droplet dilutes in the oil solution. If, on the contrary, the droplet disappears, then the wave vanishes.

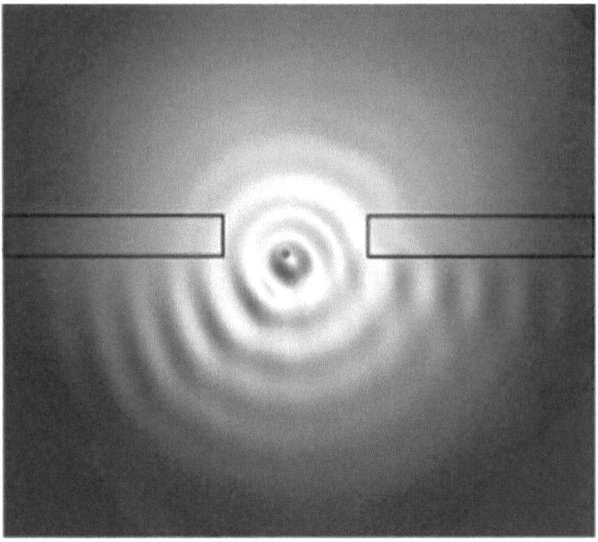

Fig. 5.19 Schematic view of the diffractive slit. The droplet and its associated wave crosses the slit at the same time (from [10])

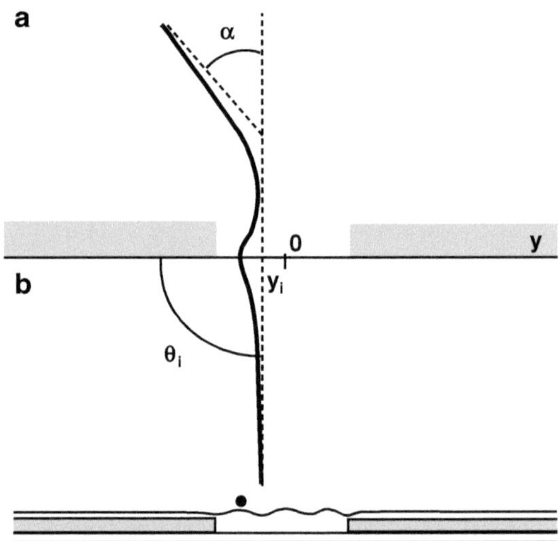

Fig. 5.20 Example of trajectory followed by a droplet (from [10])

Let us imagine that individual droplets "walk" and cross a slit (Fig. 5.19). When passing through the slit, both corpuscular and wave aspects remain. Going into further detail, the droplet trajectories can be followed individually (Fig. 5.20).

A long way before and after the slit, the motion, identified by angles θ_i et α, is straight. In the slit region, in contrast, the droplet follows a complex path. At the same time, it is seen that the wave is distorted when passing through the slit. For a certain number of individual trajectories, the number of droplets is determined as a function of the deviation angle α. Despite the chaotic behavior of the trajectories,

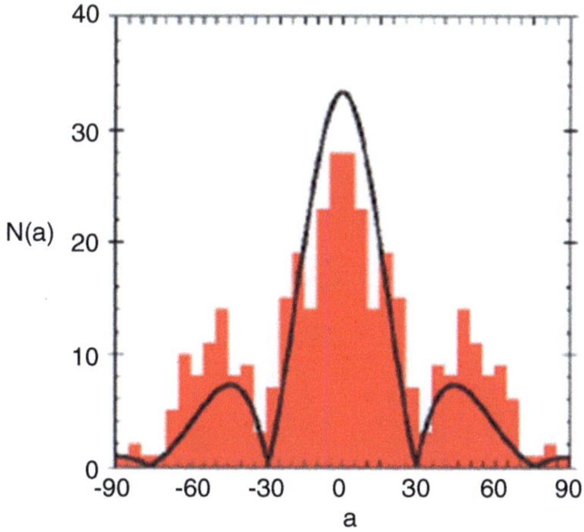

Fig. 5.21 Number of detected droplets (*in red*) as a function of the deviation angle. The black curve is the result of the simulation (from [10])

the number of droplets follows a law that clearly shows the wave character of the droplets, since a diffraction figure is obtained (Fig. 5.21) similar to that we would observe with light. In other words, determinism is not lost, since each droplet can be precisely followed along its entire trajectory.

A similar experiment was then performed using two slits instead of one. The result consists of series of maxima and minima caused by interferences.

The corpuscular-wave association gave us the idea to build a model resembling the one described above. Since the probability amplitude from quantum mechanics did not give the expected results, a semi-classical model was constructed, using the corpuscular and wave nature of an electron. In contrast with the above model, however, this one is first based on the corpuscular aspect and then on the wave behavior of the electron, meaning that each aspect is independently treated.

5.2.2 *Basis of the Model*

5.2.2.1 Corpuscular Aspect

Basis of the Calculation

The model is based on the resolution of Hamilton equations for proton and electron trajectories (Fig. 5.22). Coupled differential equations are resolved at each time, using the fourth order Runge Kutta method with an adaptive step [2]. To determine the final intensity, i.e., the number of detected electrons as a function of the detection angle, five millions of trajectories are calculated.

Fig. 5.22 Forces applied on protons and electrons

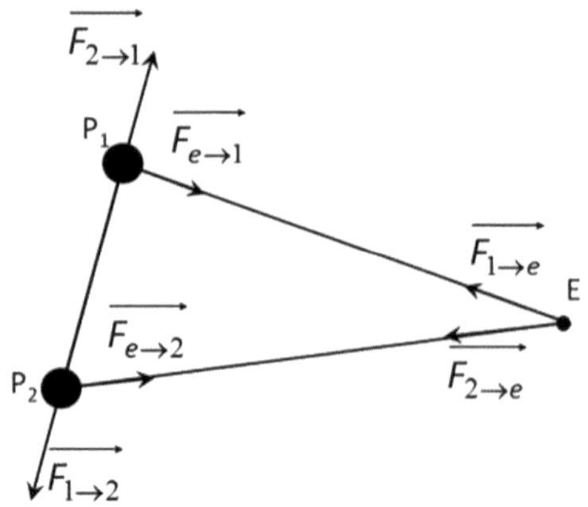

Fig. 5.23 Initial energy distribution of the emitted electrons

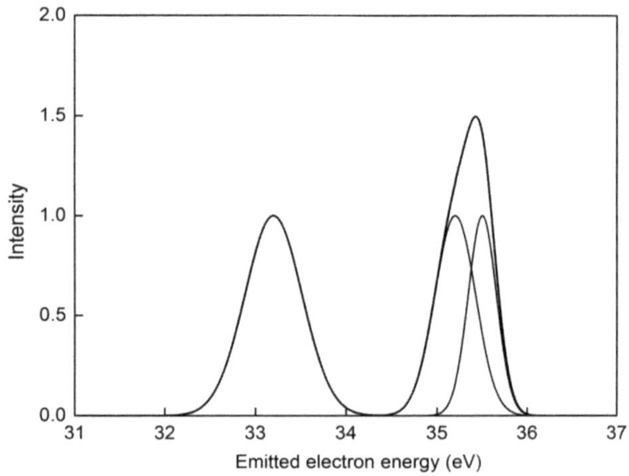

Initial Conditions

- Conditions for the electron:
  The three $2s^2\ ^1S$, $2p^2\ ^1D$ and $2s2p\ ^1P$ configurations are accounted for in the calculation. The initial total energy distribution is supposed to be a sum of three Gaussian curves (Fig. 5.23).
  As stated in Sect. 5.1.3.2, the impact parameter is of the order of 0.5 a.u. This value was used in the present calculation. The direction and sense of the electron velocity are chosen randomly. Finally, weighting is performed to take into account the exponential decrease of Auger emission (Fig. 5.24).
- Conditions for the molecule:
  Before electron capture, the molecule is randomly oriented, and the distance between the protons is the equilibrium distance $d_{HH}^{o} = 1.42$ u.a. When the Auger electron is emitted, both protons are already separated due to Coulomb

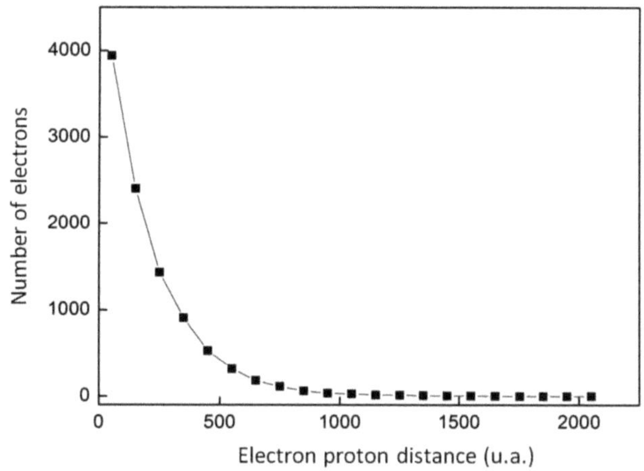

Fig. 5.24 Exponential variation of the number of emitted electrons as a function of the distance between the electron and proton centre of mass

repulsion, and the distance d_{HH} between the protons is given by relation (5.2). In addition, the protons accelerate after the capture and have a velocity $v_{HH}^{o} = d_{HH}/t_e$, where t_e is the electron emission time.

Examples of Trajectories

In order to validate the program, several tests were realized. Figures 5.25, 5.26, and 5.27 show some trajectories in particular cases. In the simplest case (Fig. 5.25), the electron (in green), has an initial velocity along (Oz). The two protons are initially in a plane perpendicular to (Oz) axis. It can clearly be seen that the electron is not deviated from its initial trajectory.

In the extreme case where the electron is far away from the protons (Fig. 5.26), the electron trajectory is linear.

Finally, when both protons have any direction, and the electron passes through them, we obtain some trajectory or other (Fig. 5.27).

5.2.2.2 Wave Aspect

When all the trajectories have been calculated, the wave aspect of the electron is taken into account. First, a sort is done depending on the detection angle. Suppose now that N trajectories are such as the detection angle is equal to θ_d with an aperture of $\Delta\theta_d = 2°$. Since the electrons are randomly emitted, they do not reach the detector at the same time. To simplify, consider two trajectories (Fig. 5.28). Both electrons are initially located on a circle whose radius is the average impact parameter, of the order of 0.5 a.u., cutting (Oz) in z_{1min} and z_{2min}, with $|z_{1\,min}| > |z_{2\,min}|$.

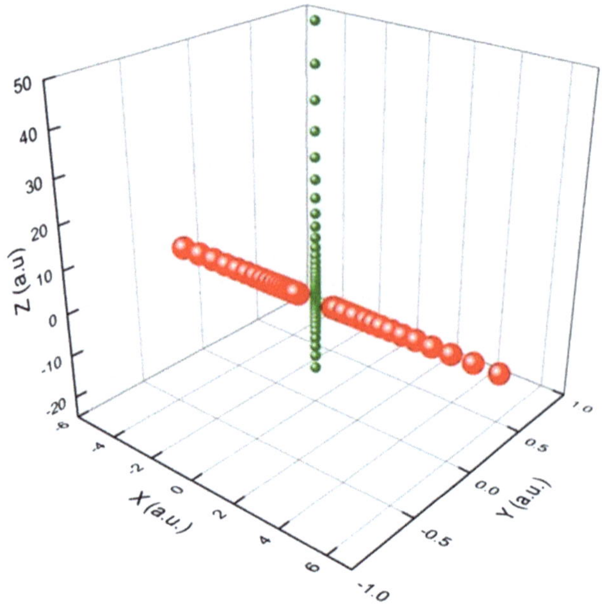

Fig. 5.25 The electron (*in green*), passes by the molecule center of masse. The molecule axis is perpendicular to the projectile trajectory

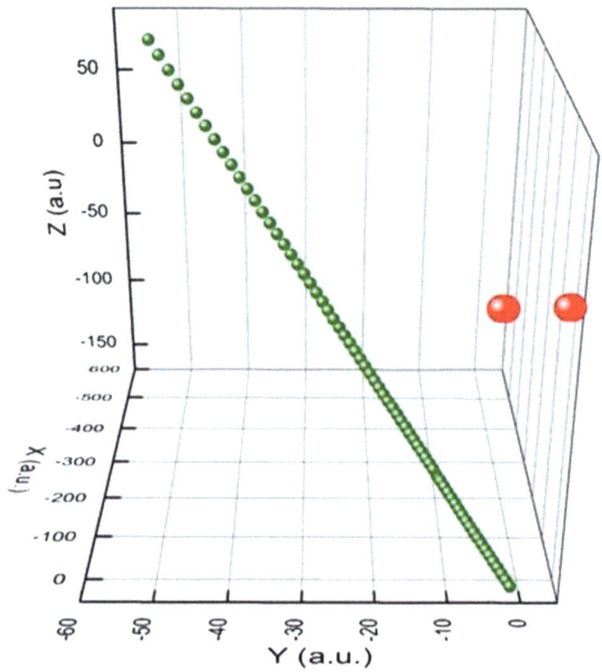

Fig. 5.26 The electron is far enough from the protons, so that its trajectory is linear

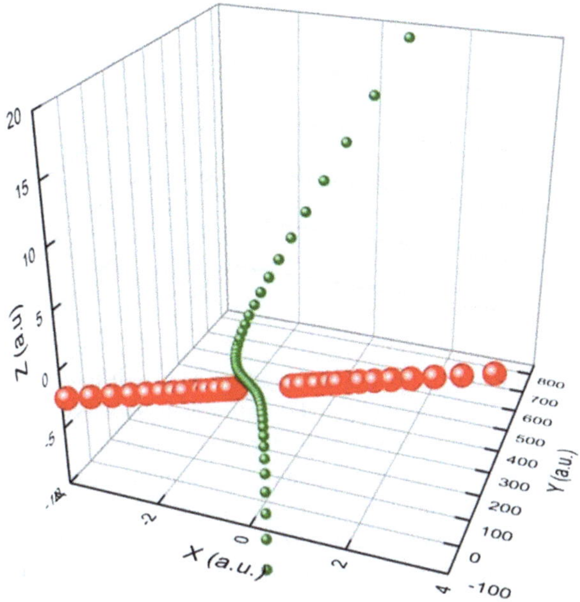

Fig. 5.27 An electron trajectory when it passes through the randomly oriented molecule

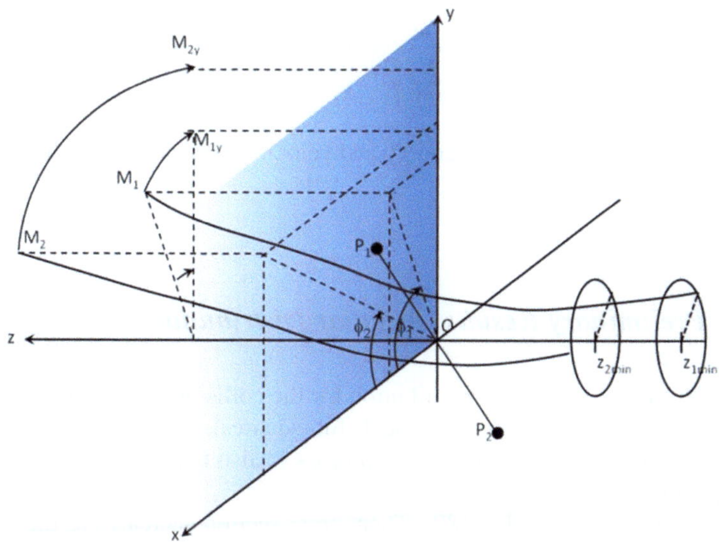

Fig. 5.28 Schematic 3D view of two electron trajectories

At the end of integration time, the two electrons are in M_1 and M_2. As indicated in Fig. 5.28, projections of these two points on xOy make angles φ_1 and φ_2 with the Ox axis, respectively. To increase the statistics in the calculation, the final points M_1 and M_2 are rotated in such a way that the problem is confined to the (yOz) plan. The points obtained after rotation are called M_{1y} and M_{2y} (Fig. 5.29).

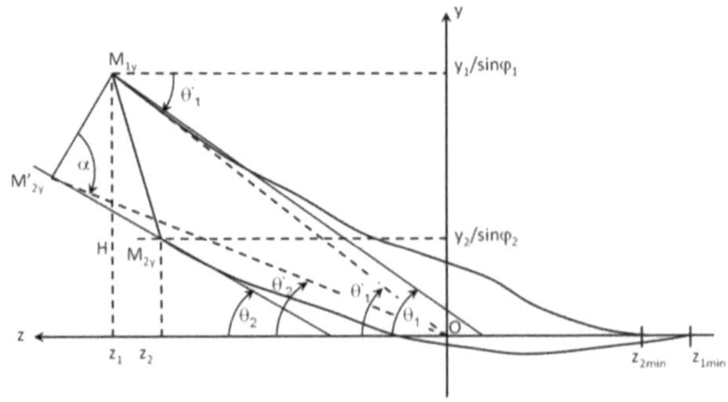

Fig. 5.29 2D view of the electron trajectories. Points M_1 et M_2 have been rotated in order to be located in (yOz) plan

The electron that is detected on M_{1y} will have reached the detector before the electron detected on M_{2y}. The latter is thus delayed compared with the first, and the missing distance (Fig. 5.29) for the electron to reach the detector is $M_{2y}M'_{2y}$.

To be sure that, at the end of the calculation, a straight line trajectory is obtained, the angles θ_I (angle between the velocity vector and (Oz)) and θ'_I (angle between the position vector and (Oz)) are compared. The calculation is ended when the relative uncertainty between the two angles is less than 1 %.

Finally, the total amplitude is calculated, at a fixed angle, using the following expression

$$A = \sum_{k=0}^{N} e^{-i\frac{E_k}{\hbar}\left(t_{k\max} - \frac{M_{iy}M'_{iy}}{v_k}\right)} \tag{5.22}$$

where E_k and v_k are the electron energy and velocity, respectively, at the end of the collision.

5.2.3 *Preliminary Results: Angular Distribution*

Figure 5.30 shows the angular distribution for the collision $He^{2+} + H_2$ collisions at 30 keV, obtained with the present model (full squares). In this model, the interaction between the projectile and the two protons is also taken into account, and the impact parameter is 0.

A maximum is observed at 180°. Then, cross section decreases as the detection angle decreases. In the range from 90° to 170°, series of maxima and minima are clearly visible. Experimental maxima and minima, indicated by red arrows in Fig. 5.30, are very close to the calculated ones. In conclusion, this semi-classical model is able to reproduce the oscillations.

At this point, we can pursue the discussion of these experiments in greater depth. In the case of photon interference experiments, it is now widely accepted that the corpuscular and wave aspects of light manifest themselves as follows: When the

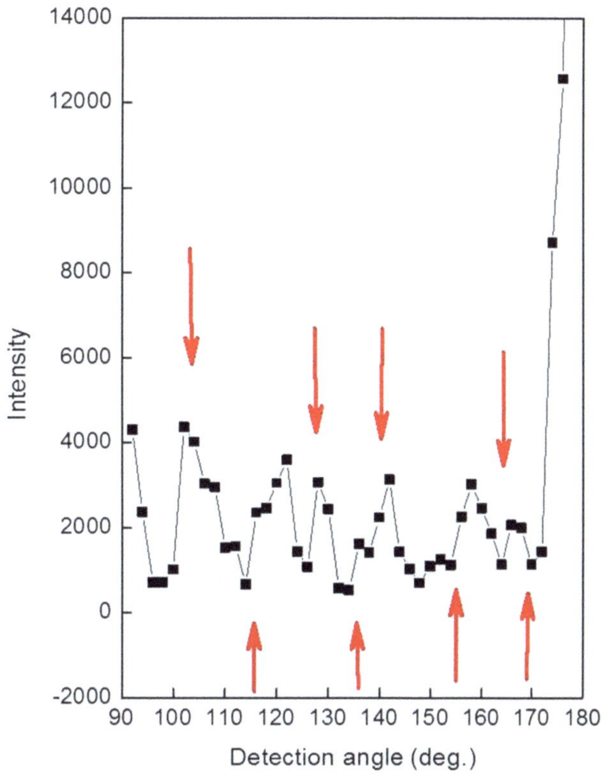

Fig. 5.30 Typical calculated angular distribution obtained for $He^{2+} + H_2$ collisions at 30 keV (*full squares*). The *vertical arrows* indicate the position of the experimental minima and maxima

light crosses the slits, it takes its wave behavior, giving rise to interferences. Thus, when light reaches the detector, it recovers its corpuscular behavior, because the wave function is supposed to collapse.

The present calculation can lead to another possible interpretation. From the time when the electron is emitted from the projectile, it maintains its corpuscular behavior. Its wave behavior, which can be present also during the collision of the electron with the protons, occurs only in the determination of the phase shift. In this sense, the present calculation is very similar to what happens in the droplet interference experiment described above. Of course, this interpretation is at present hypothetic, and will need to be analyzed in more detail. This model will now be applied to other collision systems, at different projectile velocities. In particular, energy distributions of emitted electrons at a fixed angle will be calculated and compared with experiments.

References

1. Barker, R.B., Berry, H.W.: Electron energy distribution from ionizing collisions of Helium and Neon ions with Helium. Phys. Rev. **151**, 14 (1966)
2. Devdariani, A.Z., et al.: Electron spectra from the autoionization of quasimolecules, Zh. Eksp. Teor. Fiz. **73,** 412 (1977) [Sov. Phys. JEPT. **46,** 215 (1977)]
3. van der Straten, P., Morgenstern, R.: Interference of autoionising transitions in fast ion-atom collisions. J. Phys. B. **18**, 1361 (1986)
4. Barrachina, R.O., Macek, J.H.: Theory of emission-angle-dependent Auger transitions in ion-atom collisions. J. Phys. B. **22**, 2151 (1989)
5. Otranto, S., et al.: Autoionizing states of He: projectile velocity-dependent lifetime. Phys. Rev. A. **63**, 022713 (2001)
6. Barrachina, R.O.: Post-collision interaction effects in the autoionization of atoms excited by ion impact. Nucl. Intr. Meth. **132**, 288 (1997)
7. Caputi, K.I.: Efectos postcolisionales en procesos de emissión Auger, trabajo especial carrera de licenciatura en física, San Carlos de Bariloche (1998)
8. Jain, A., et al.: State-selective double-electron capture in He^{2+} + He collisions at intermediate impact energies. Phys. Rev. A. **39**, 1741 (1989)
9. Barrachina, R.O., Zitnik, M.: Young's interference effect in the autoionization of atoms colliding with molecules. J. Phys. B. **37**, 3847 (2004)
10. Couder, Y., Fort, E.: Single-particle diffraction and interference at a macroscopic scale. Phys. Rev. Lett. **97**, 154101 (2006)

Appendix 1: System of Atomic Units

International System of Units

The International System of Units (SI: *Système International d' unités*), based on the meter, the kilogram, and the second (MKS), has been officially adopted by almost every nation in the world and is, by far, the most widely preferred system in Science. Thus, in principle, one would be tempted to use international units as a common practice in Physics. However, it is clear that the units that make up the SI have been defined to fit our human scale. This is obviously true for the meter, the kilogram, the second, and even the candela (cd), which is defined as being roughly equal to the luminous intensity of a common candle. However, when we have to deal with things that are beyond the human scale, the SI is far from being the most convenient choice of units. Of course, we can resort to multiples or fractions of the basic units that would be better suited to describing the physical system at hand. In practice, at the atomic scale, everything is too small, too light or too fast to be described in meters, kilograms or seconds. Therefore, suitable fractions would be required, as for instance the nanometer (1 nm $= 10^{-9}$ m), the yoctogram (1 yg $= 10^{-27}$ kg) or the femtosecond (1 fm $= 10^{-15}$ s). Thus we would be using the femtoJoule (1 fJ $= 10^{-15}$ kg m^2/s^2) for energy, or the petaHertz (1 pHz $= 10^{15}$ s^{-1}) for frequency, and expressing velocities in nanometers per femtosecond.

Another issue that could be more problematic is that of complicated combinations of physical constants. For instance, the ionization energy of the hydrogen atom would be

$$I = \frac{m_e e^4}{32\,\pi^2\,\hbar^2\,\varepsilon_o^2} \tag{A.1}$$

where

$$m_e = 9.1093826 \times 10^{-31}\mathrm{kg} \tag{A.2}$$

F. Frémont, *Young-Type Interferences with Electrons*, Springer Series on Atomic, Optical, and Plasma Physics 77, DOI 10.1007/978-3-642-38479-0,

and

$$e = 1.60217653 \times 10^{-19}\mathrm{C} \tag{A.3}$$

are the mass and the charge of the electron, respectively,

$$\hbar = 1.05457168 \times 10^{-34}\ \mathrm{kg\,m^2\,s^{-1}} \tag{A.4}$$

is the reduced Planck's constant, and

$$\varepsilon_o = 8.85418781762 \times 10^{-12}\ \mathrm{A^2\,s^4\,kg^{-1}\,m^{-1}} \tag{A.5}$$

is the vacuum permittivity.

To reduce these difficulties in calculation, atomic physics adopted the simple system of atomic units. This system can be related to the Bohr model and is based on the simplification of certain constants. More precisely, we set $e = 1$, $\hbar = 1$, $1/(4\,\pi\,\varepsilon_o) = 1$ and $m_e = 1$.

Atomic Units

Let us consider an electron in circular motion around a proton (Fig. A.1). The electron is bound by a central force

$$\vec{F} = \frac{-e^2}{4\,\pi\,\varepsilon_o}\frac{\vec{u_r}}{r^2} = -m_e\frac{v^2}{r}\vec{u_r} \tag{A.6}$$

where r is the classical distance between the electron and the proton, and v is the velocity of the electron.

The electron momentum ℓ is quantified, so that

$$\ell = n\hbar = m_e v r \tag{A.7}$$

where n is the principal quantum number.

The total electron energy is given by

$$E = \frac{1}{2}m_e v^2 - \frac{e^2}{4\,\pi\,\varepsilon_o}\frac{1}{r} \tag{A.8}$$

Using the previously described atomic unit system, the above three relations become

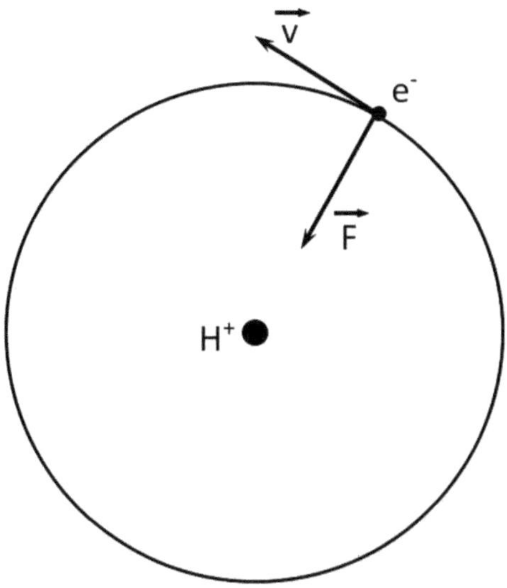

Fig. A.1 Classical representation of hydrogen. A proton (H^+) attracts an electron (e^-) via a central force $\overrightarrow{F}$. The electron has a constant velocity $\overrightarrow{v}$

Table A.1 Equivalent in atomic units for the most usual quantities

Quantities	Equivalent for 1 a.u.
Length	0.53 nm
Energy	27.21eV
Time	2.4×10^{-17} s
Velocity	2.2×10^{6} m/s
Mass	9.1×10^{-31} kg
Charge	1.6×10^{-19} C
Angular momentum	1.05×10^{-34} Js

$$\begin{cases} -\frac{1}{r^2} = -\frac{v^2}{r} \\ n = vr \\ E = \frac{1}{2}v^2 - \frac{1}{r} \end{cases} . \tag{A.9}$$

From the last relations, the total energy of the electron on the fundamental orbital $n = 1$ is $E = -0.5$ a.u. ≈ -13.605 eV. The distance in a.u. is the Bohr radius, thus 1 a.u. $\approx$ 0.53 nm. Finally, all the quantities in a.u., such as the cross section, are derived from the previous quantities (Table A.1).

Appendix 2: Scattering Cross Section

Definition of the Scattering Cross Section

The number dN_d of scattered particles per second detected in the solid angle $d\Omega$ is proportional to the length $v\,dt$ of the collision, and proportional to the number n_p of incident particles per volume unit (Fig. A.2).

The coefficient of proportionality has the dimension of a surface. It can be given as

$$dN_d = n_p v dt d\sigma \tag{A.10}$$

where $d\sigma$ is the differential cross section. Thus the differential scattering cross section in solid angle is:

$$\frac{d\sigma}{d\Omega} = \frac{dN_d}{n_p v dt d\Omega} \tag{A.11}$$

Scattering Cross Section as a Function of the Impact Parameter

The solid angle is connected, classically, to the impact parameter b (Fig. A.3). The number of scattered particles in the solid angle $d\Omega$ is the number of particles that were, before the collision, in a circular ring, with radii in the range from b to $b + db$. This leads to

$$dN_d = n_p v dt 2\pi\, b db \tag{A.12}$$

F. Frémont, *Young-Type Interferences with Electrons*, Springer Series on Atomic, Optical, and Plasma Physics 77, DOI 10.1007/978-3-642-38479-0,

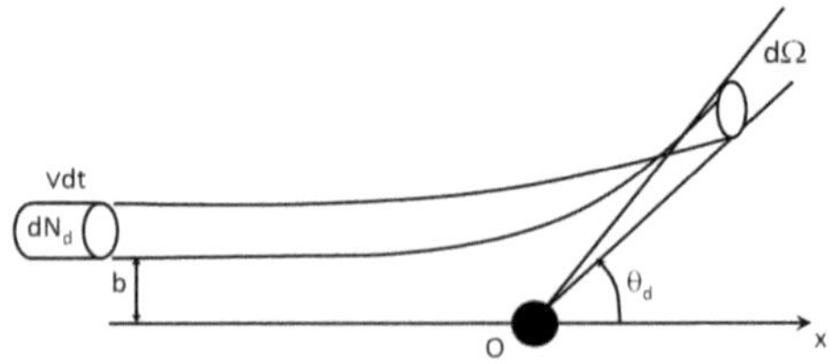

Fig. A.2 Classical representation of particle trajectory disturbed by the presence of an atom. The particles contained in a volume *vdtdS* are detected, after the collision, in a solid angle $d\Omega$

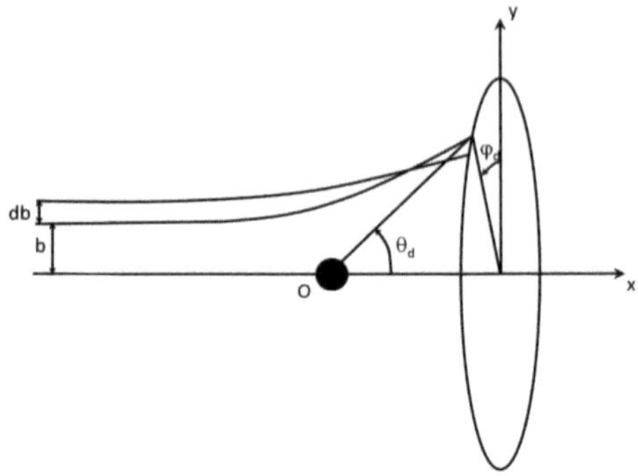

Fig. A.3 Connection between the impact parameter b and the final scattering angle θ_d

where $2\pi\, bdb$ is the surface of the circular ring. Combining Eqs. A.11 and A.12, we obtain

$$\frac{d\sigma}{d\Omega} = 2\pi\, b \frac{db}{d\Omega} \tag{A.13}$$

Replacing $d\Omega$ by $2\pi\ \sin\theta_d\, d\theta_d$, the cross section is

$$\frac{d\sigma}{d\Omega} = \frac{b}{\sin\theta_d}\left|\frac{db}{d\theta_d}\right| \tag{A.14}$$

The absolute value is only here to ensure that the cross section is positive.

Appendix 3: Interference Term in H_2 Ionization

Ionization Amplitude of Probability

Consider a system described by a time-independent Hamiltonian H_o. The eigenstates of the system form a complete basis set $\{|\varphi_n\rangle\}$ and are solutions of the time-independent Schrödinger equation

$$H_o\,|\varphi_n\rangle = E_n\,|\varphi_n\rangle \tag{A.15}$$

where E_n is the eigenvalue associated with a state $|\varphi_n\rangle$.

The system is supposed to be disturbed by a small time-dependent perturbation W

$$H(t) = H_o + W(t) \tag{A.16}$$

with $W \ll H_o$. Consequently, the $|\varphi_n\rangle$ are no longer eigenstates of $H(t)$. We can now write $W(t) = \alpha\, w(t)$, where $\alpha \ll 1$. The new eigenstates of $H(t)$ are $|\psi_n(t)\rangle$ and verify

$$H(t)\,|\psi_n(t)\rangle = H_o\,|\psi_n(t)\rangle + \alpha\,w(t)\,|\psi_n(t)\rangle = i\,\hbar\,\frac{d|\psi_n(t)\rangle}{dt} \tag{A.17}$$

At any time, the $|\psi_n(t)\rangle$ can be decomposed on the $|\varphi_n\rangle$ basis set.

$$|\psi_n(t)\rangle = \sum_n b_n(t)e^{-i\frac{E_n}{\hbar}t}\,|\varphi_n\rangle \tag{A.18}$$

Replacing Eq. A.18 in relation (A.17), we obtain

$$i\hbar\sum_n \frac{d}{dt}\left(b_n(t)e^{-i\frac{E_n}{\hbar}t}\right)|\varphi_n\rangle = \sum_n b_n(t)e^{-i\frac{E_n}{\hbar}t}(H_o + \alpha\,w(t)\,)|\varphi_n\rangle \tag{A.19}$$

F. Frémont, *Young-Type Interferences with Electrons*, Springer Series on Atomic, Optical, and Plasma Physics 77, DOI 10.1007/978-3-642-38479-0,

The projection of the last equation on $\langle\varphi_p|$ gives rise to a system of coupled differential equations

$$i\hbar\frac{db_p(t)}{dt} = \alpha\sum_n b_n(t)\,e^{-i\frac{\Delta E_{np}}{\hbar}t}\,w_{pn}(t) \tag{A.20}$$

In this relation, $w_{pn}(t) = \langle\varphi_p|w(t)|\varphi_n\rangle$ and $\Delta E_{np} = E_p - E_n$.

Since the perturbation is small in comparison to the total Hamiltonian, the coefficients $b_p(t)$ are developed in terms of the power of α

$$b_p(t) = b_p^{(o)}(t) + \alpha\, b_p^{(1)}(t) + +\alpha^2\, b_p^{(2)}(t) + \ldots \tag{A.21}$$

At zeroth order, $\alpha = 0$, so that $db_p^{(o)}(t)/dt = 0$. Thus, the system remains in its initial state. At first order,

$$i\hbar\frac{db_p^{(1)}}{dt} = \alpha\sum_n b_n^{(o)}(t)\,e^{-i\frac{\Delta E_{np}}{\hbar}t}\,w_{pn}(t) \tag{A.22}$$

To simplify, suppose that the system is in an initial state $|\varphi_i\rangle$, with an initial energy E_i.

$$i\hbar\frac{db_p^{(1)}}{dt} = \alpha\, b_i^{(o)}(t)\,e^{-i\frac{\Delta E_{ip}}{\hbar}t}\,w_{pi}(t) \tag{A.23}$$

with $b_i^{(o)}(t) = \text{cte} = b_i^{(o)}(t = 0) = 1$. Equation A.23 leads to

$$b_p^{(1)} = \frac{1}{i\hbar}\int_0^t e^{-i\frac{\Delta E_{ip}}{\hbar}t'}\,W_{pi}(t')\,dt' \tag{A.24}$$

Finally, it can be seen that the amplitude of probability due to the perturbation is the Fourier transform of the matrix element $W_{pi}(t')$.

Fermi's Golden Rule

When the perturbation turns on at $t = 0$ and is almost constant over a time interval $[0, t]$, the amplitude can be derived analytically

$$b_f^{(1)} = \frac{1}{i\hbar}\,W_{if}(t)\,e^{-i\frac{\Delta E_{ip}}{\hbar}t}\,\frac{\sin\left(\frac{\Delta E_{ip}}{\hbar}t\right)}{\frac{\Delta E_{ip}}{\hbar}} \tag{A.25}$$

The transition probability is thus, in atomic units

$$P_{if}(t) = \left|W_{if}(t)\right|^2 \frac{\sin^2\left(\Delta E_{ip}t\right)}{\left(\Delta E_{ip}\right)^2} \tag{A.26}$$

When an atom, initially in its fundamental state, is ionized, there is a coupling between the initial state and the continuum. An interval of energies dE_f of the continuum is populated by a number of states $dn = \rho\,(E_f)\,dE_f$, where $\rho\,(E_f)$ is the density of states.

Suppose that W and $\rho\,(dE_f)$ are smoothly varying functions of energy near the states in the continuum. The total ionization probability per time unit is the sum over all the individual ionization states

$$p_{if}(t) = \frac{1}{t} \sum_{f'\ near\ f} P_{if}(t) \tag{A.27}$$

Since the states in the continuum are close to each other, the discrete sum can be replaced by an integral

$$p_{if}(t) \approx 4\rho(E_f)\left|W_{if}\right|^2 \int_0^\infty \frac{1}{t} \frac{\sin^2\left(\Delta E_{ip}t\right)}{\left(\Delta E_{ip}\right)^2} \tag{A.28}$$

The integral is equal to $\sqrt{\pi/2}$. Finally, the ionization probability is proportional to

$$p_{if}(t) \approx 4\sqrt{\pi/2}\rho(E_f)\left|W_{if}\right|^2 \tag{A.29}$$

This expression is known as Fermi's golden rule.

Application to H_2 Ionization

Many models describe the ionization of an atom or a molecule after ion impact. The goal here is not to give any details on these models, but only to obtain evidence of the interference term. To do this easily, we use the Born approximation. In the framework of quantum mechanics, the first-order Born approximation describes the heavy particle as a plane wave before and after the collision. The electron is described by bound and continuum wave functions before and after the collision, respectively. Consequently, assuming that the continuum wave function is a plane wave, both the initial and final states can be written

$$\Phi_i \approx \frac{1}{N_i} e^{i\overrightarrow{K_i}.\overrightarrow{R}} [\varphi_A(r_A) + \varphi_B(r_B)]$$
$$\Phi_f \approx \frac{1}{N_f} e^{i\overrightarrow{K_f}.\overrightarrow{R}} e^{ikr} \tag{A.30}$$

In these expressions, N_i and N_f are normalization factors, $\overrightarrow{K_i}$ and $\overrightarrow{K_f}$ are the initial and final momenta, respectively, of the projectile, $\overrightarrow{R}$ is the distance between the projectile and the center of mass of the target, and $\overrightarrow{r}$ is the distance between the electron and the center of mass of the target.

Using Eq. A.24, the amplitude is

$$b_{if}(t) \propto \int e^{-i\overrightarrow{K_f}.\overrightarrow{R}} e^{-ikr} e^{i\overrightarrow{K_i}.\overrightarrow{R}} (V(r_A)\varphi_A(r_A) + V(r_B)\varphi_B(r_B))d^3r$$

Replacing $V(r_A)$ and $V(r_B)$ by $-1/r_A$ and $-1/r_B$, respectively, and assuming that $\varphi_{A,B}(r_{A,B}) \propto e^{-\alpha r_{A,B}}$, with $\overrightarrow{r_{A,B}} = \overrightarrow{r} \pm \overrightarrow{r_H}$ where $\overrightarrow{r_H}$ is the vector defining the position of one proton referred to the center of mass of the molecule

$b_{if}(t) \propto \int e^{i\Delta\overrightarrow{K}.\overrightarrow{R}} e^{-ikr}\left(\frac{1}{r_A}e^{-\alpha r_A} + \frac{1}{r_B}e^{-\alpha r_B}\right)d^3r$ where $\Delta\overrightarrow{K} = \overrightarrow{K_i} - \overrightarrow{K_f}$ is the momentum transfer. The integral can be separated into two terms

$$b_{if}(t) \propto \int e^{i\Delta\overrightarrow{K}.\overrightarrow{R}} e^{-ikr} \frac{1}{r_A} e^{-\alpha r_A} d^3r + \int e^{i\Delta\overrightarrow{K}.\overrightarrow{R}} e^{-ikr} \frac{1}{r_B} e^{-\alpha r_B} d^3r$$

$$b_{if}(t) \propto \int e^{i\Delta\overrightarrow{K}.\overrightarrow{R}} e^{-ik\left|\overrightarrow{r_A} - \overrightarrow{r_H}\right|} \frac{1}{r_A} e^{-\alpha r_A} d^3r + \int e^{i\Delta\overrightarrow{K}.\overrightarrow{R}} e^{-ik\left|\overrightarrow{r_B} + \overrightarrow{r_H}\right|} \frac{1}{r_B} e^{-\alpha r_B} d^3r$$

Since $r_H \ll r_{A,B}$, $e^{-ik\left|\overrightarrow{r_{A,B}} \mp \overrightarrow{r_H}\right|} \approx e^{-ikr_{A,B}} e^{ikr_H\cos\theta}$, where θ defines the orientation of the molecule with respect to a fixed axis, the amplitude of probability is therefore

$$b_{if}(t) \propto \int e^{i\Delta\overrightarrow{K}.\overrightarrow{R}} e^{-ikr_A} e^{ikr_H\cos\theta} \frac{1}{r_A} e^{-\alpha r_A} d^3r + \int e^{i\Delta\overrightarrow{K}.\overrightarrow{R}} e^{-ikr_B} e^{-ikr_H\cos\theta} \frac{1}{r_B} e^{-\alpha r_B} d^3r$$

$$b_{if}(t) \propto e^{ikr_H\cos\theta} \int e^{i\Delta\overrightarrow{K}.\overrightarrow{R}} e^{-ikr_A} \frac{1}{r_A} e^{-\alpha r_A} d^3r$$
$$+ e^{-ikr_H\cos\theta} \int e^{i\Delta\overrightarrow{K}.\overrightarrow{R}} e^{-ikr_B} \frac{1}{r_B} e^{-\alpha r_B} d^3r$$

To simplify, we now assume that the integrals in the latter expression are equal. Thus

$$b_{if}(t) \propto A\left(e^{ikr_H\cos\theta} + e^{-ikr_H\cos\theta}\right)$$

The amplitude of probability is finally

$$b_{if}(t) \propto \cos\left(\overrightarrow{k}.\overrightarrow{r_H}\right) \tag{A.31}$$

The latter result shows that oscillation occurs, due to the initial state of the molecule, defined as a linear combination of two atomic states.

Appendix 4: Post Collision Interaction in Atomic Collisions

The Principle of Post Collision Interaction

It has been an acknowledged fact for many decades now that fast energy collisions on multielectronic targets, such as He, Ne, or Ar, for example, can give rise to electron energy spectra that consist of two different structures: (i) a continuous and decreasing background due to direct ionization

$$A^+ + B \rightarrow A^+ + B^+ + e^- \tag{A.32}$$

and (ii), superimposed onto this background, well-defined peaks, caused by the Auger effect (emission of an electron e_A^-) after double or multiple excitation of the target

$$\begin{aligned} A^+ + B &\rightarrow A^+ + B^{**} \\ B^{**} &\rightarrow B^+ + e_A^- \end{aligned} \tag{A.33}$$

At high projectile energies, Auger spectra generally consist of several peaks. In the case of double excitation of He, for example, these peaks are associated with the deexcitation of $2\ell 2\ell'$ configurations (32–36 eV) and $2\ell n\ell'$, with $\mathrm{n} > 2$ configurations (36–42eV), as shown in Fig. A.4 [1] for the collision system $\mathrm{H}_2^+ + \mathrm{He}$.

The full width at half maximum (FWHM) of the peaks was found to be ~0.3eV for the structures of highest intensity. Analysis of the double excitation of He after impact of slow ions [2] also revealed similar structures (Fig. A.5). Nevertheless, the width of the peaks was found to be higher than that found at high velocities by a factor of at least 3.

[1] Rudd, M.E.: Observation of autoionization levels in He by positive ion bombardment. Phys. Rev. Lett. **13**, 503 (1964)

[2] Berry, H.W.: Energy distribution of electrons from ionizing collision of heavy particles. Phys. Rev. **121,** 1714 (1961)

F. Frémont, *Young-Type Interferences with Electrons*, Springer Series on Atomic, Optical, and Plasma Physics 77, DOI 10.1007/978-3-642-38479-0,

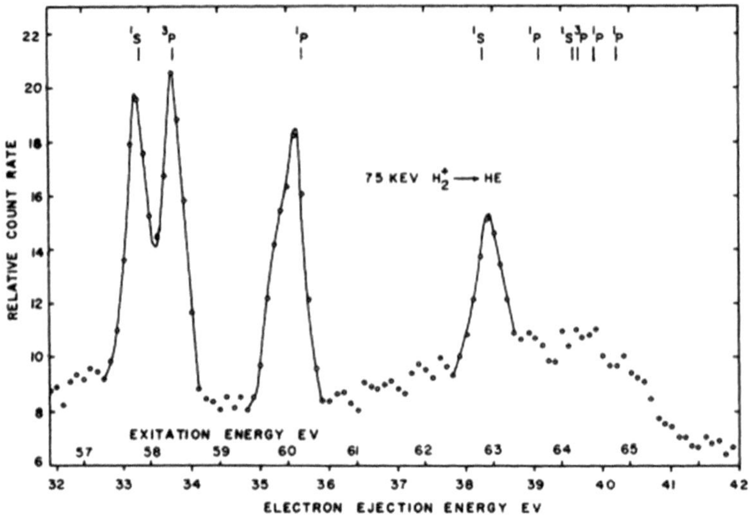

Fig. A.4 Example of Auger energy spectra in the case of fast ion impact on He

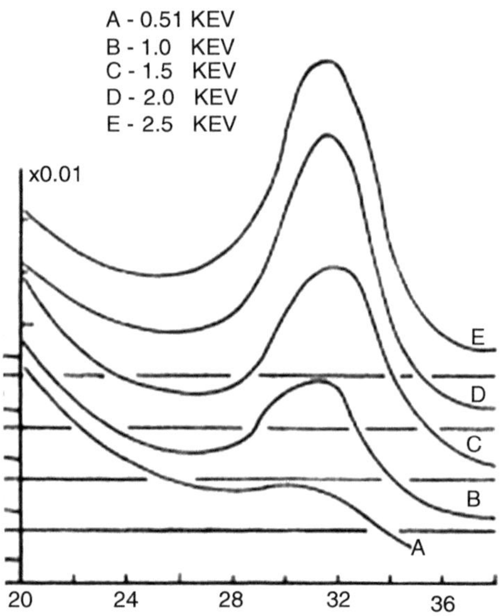

Fig. A.5 Auger electron spectra obtained in He^+ + He collisions at projectile energies in the range 0.51–2.5 keV

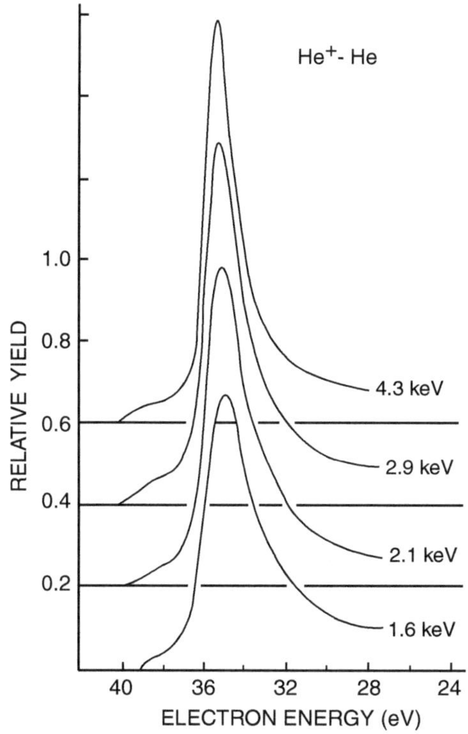

Fig. A.6 Auger electron spectra obtained in He^+ + He collisions at projectile energies in the range 1.6–4.3 keV

Also, as seen in Fig. A.5, the width increases with decreasing projectile energy. In 1966, the He^+ + He system was experimentally reexamined to ensure that this increase was not accidental, or due to experimental artifacts [3]. However, as seen in Fig. A.6, the width is again of the order of 1 eV, as in the previous experiment. In addition, careful analysis showed a shift in the energy maximum to lower energies when the projectile energy E_p decreased.

Quantitatively, the authors demonstrated that the energy shift and the width had a "linear dependence on the inverse square root of the ion energy", i.e. its velocity. This is illustrated in Fig. A.7 for both experimental quantities at low energies (full squares and circles) and at high energy (open square and circle). The experimental results are well fitted using a function of the type $a + b/\sqrt{E_p}$.

According to the authors, "there is no mechanism inherent in the autoionization process which would account for the shift in the peak energy with ion velocity" [3]. The authors rather suggested that what was observed was a consequence of autoionization in the field of the target ion. Gradually as the projectile velocity decreased, the electron was increasingly influenced by the target, due to Coulomb interaction.

[3] Barker, R.B., Berry, H.W.: Electron energy distribution from ionizing collisions of Helium and Neon ions with Helium. Phys. rev. **151**, 14 (1966)

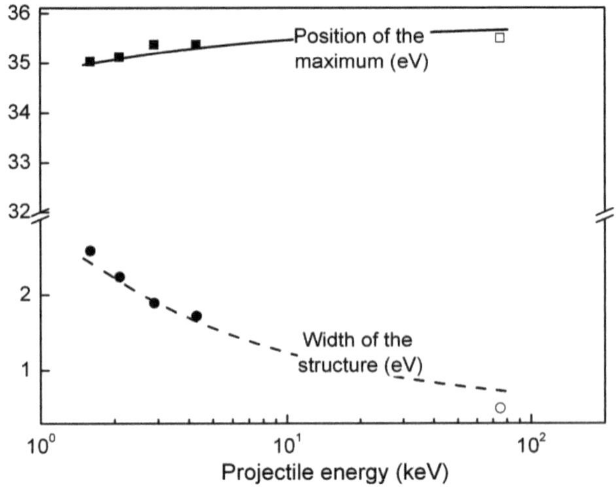

Fig. A.7 Experimental position and width of the Auger peaks in He^+ + He collisions (*full squares*). The full and *dashed curves* are the results of fitting procedures, using functions of the type $a + b/\sqrt{E_p}$

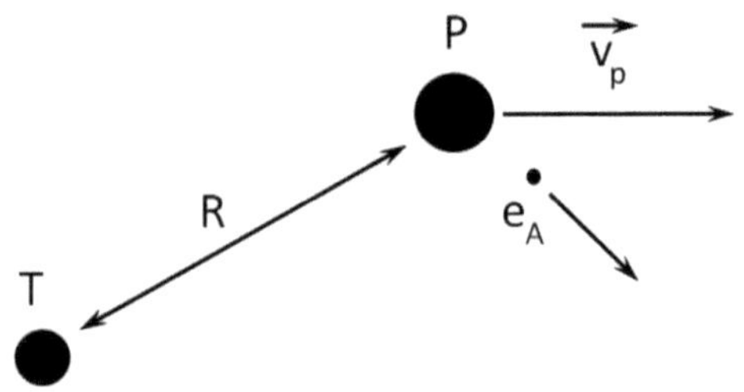

Fig. A.8 Schematic diagram of the collision. The projectile *P* emits a Auger electron, in the vicinity of the charged target

The Barker and Berry Model

Consider N_o atoms *P* with velocity v_p emitting Auger electrons of energy E_o without the presence of any ion. At time *t*, *N* electrons have been emitted, and *dN* electrons are emitted in an interval of time *dt*.

$$dN = (N_o - N)\,\Gamma dt \tag{A.34}$$

where Γ is the natural linewidth of the structure. After integration on time, the number *N* of emitted electrons is given by the relation

$$N = N_o\, e^{-\Gamma t} \tag{A.35}$$

Figure A.8 shows the collision partners at time *t*, when the projectile has emitted the electron in the field of the target. Just after the emission, the projectile is at a distance *R* from the target, approximately defined as $R = v_p\, t$, if we neglect the contribution of the impact parameter. The energy of the electron is $E - Q/R$ where *E* is the kinetic energy and $-Q/R$ is the potential energy due to the proximity of the target of charge *Q*.

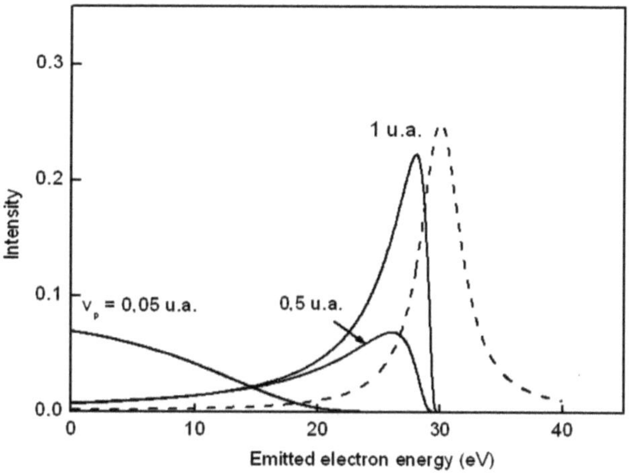

Fig. A.9 Energy distributions of Auger electrons determined using relation (A.37). *Dashed curve*: Lorentzian profile for an infinite projectile velocity; *full curves*, profiles for velocities of 1, 0.5, and 0.05 a.u

Therefore, the maximum kinetic energy difference between $t = 0$ and $t = \infty$ is

$$\Delta E_m \approx \frac{Q}{R} \approx \frac{Q}{v_p t} \approx \frac{Q\Gamma}{v_p} \tag{A.36}$$

The present result is the most important one to remember. The influence of the target is proportional to its charge, depends on the ejection time, through the lifetime Γ, and is significant at the lowest velocities.

To determine the Auger profile, we express dN/dE as a function of the emitted electron energy E. It comes, from relation (A.35)

$$\frac{dN}{dE} = \frac{QN_o\Gamma}{v_p(E - E_o)^2} e^{-\frac{Q\Gamma}{v_p(E-E_o)}} \tag{A.37}$$

where E_o is the maximum of the distribution in the absence of the target.

Figure A.9 shows several energy distributions obtained for Auger electrons of energy $E_o = 30\text{eV}$, corresponding to double excited $2s^2\ ^2S$ configuration of He. The dashed curve is a Lorentzian profile, characterizing the energy in absence of the target, or when the projectile has a high velocity. When v_p decreases, the energy distribution is strongly modified. First, the profile becomes increasingly asymmetric. A large tail is observed at low velocity. Then, the energy maximum and the width increase according to

$$\Delta E = \frac{Q\Gamma}{2v_p}$$

$$\delta E \approx 1.07 \frac{Q\Gamma}{v_p} \tag{A.38}$$

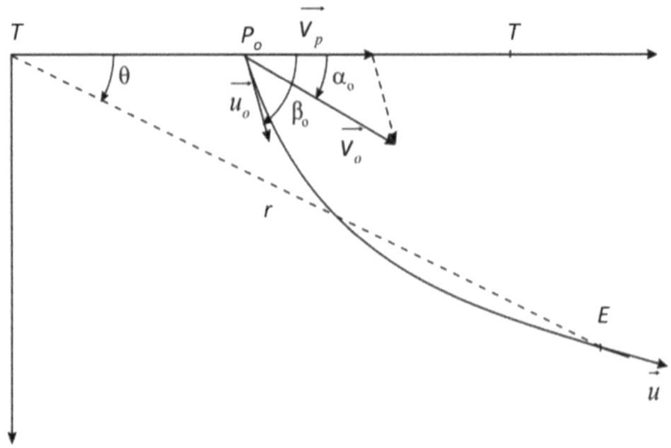

Fig. A.10 Schematic view of the trajectory of an electron E emitted from a projectile P in the field of an ionized target T. For more details, see text

This model is sufficient for determining the main features of this post-collision interaction (PCI) effect. However, it does not take into account the emission angle of the electron, which plays a determinant role.

Angular Dependence of the PCI Effect

The influence of the emission angle on PCI effects can be derived from classical kinematics. In the following model [4], the target T is at rest, and the projectile P has a constant velocity $\overrightarrow{v_p}$ along the *(Tx)* axis (Fig. A.10). The electron is emitted at a time $t_o = TP_o/v_p$. with velocities $\overrightarrow{u_o}$ and $\overrightarrow{v_o}$ in the laboratory and projectile frames, respectively. At infinity, the electron, in the same frames, has velocities $\overrightarrow{u}$ and $\overrightarrow{v}$. All along the trajectory, the electron is defined by its polar coordinates r and θ.

The kinetic energy difference between $t = 0$ and infinity is

$$\Delta E = \frac{1}{2}v^2 - \frac{1}{2}v_o^2 = \frac{1}{2}\left(\overrightarrow{u} + \overrightarrow{v_p}\right)^2 - \frac{1}{2}\left(\overrightarrow{u_o} + \overrightarrow{v_p}\right)^2$$

$$\Delta E = \frac{1}{2}\left(u^2 - u_o^2\right) + \left(\overrightarrow{u} - \overrightarrow{u_o}\right) . \overrightarrow{v_p}$$

$$\Delta E = -\frac{Q\Gamma}{v_p} + \overrightarrow{\Delta v} . \overrightarrow{v_p} \tag{A.39}$$

From the conservation of orbital momentum, and assuming small deviations of the trajectories,

[4] Van der Straten, P., Morgenstern, R.: Interference of autoionizing transitions in fast ion-atom collisions. J. Phys. B **19**, 1361 (1986)

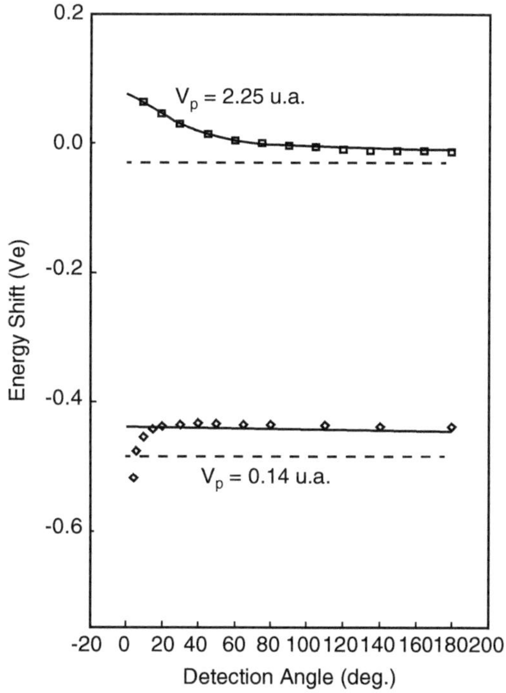

Fig. A.11 Energy shift for the collision between a singly charged ion and a Helium target, at projectile velocities of 0.14 and 2.25 a.u. *Open squares*, results of coupled equations resolution; *Full curve*, present model; *dashed lines*, Barker and Berry model

$$\Delta E = -\frac{Q\Gamma}{v_p}\left(1 - \frac{v_p}{\left|\overrightarrow{u_o} + \overrightarrow{v_p}\right|}\right) \tag{A.40}$$

This expression contains the angular dependence through $\left|\overrightarrow{u_o} + \overrightarrow{v_p}\right|$. Figure A.11 shows the energy difference in the collision between a singly charged ion and a He target, calculated by means of coupled equations (open squares), and using the relation (A.40), at projectile velocities of 2.25 and 0.14 a.u. The energy difference predicted by the Barker and Berry model is also shown (dashed lines). At the highest velocity, the present model agrees with the most precise calculations. At 0.14 a.u., the agreement is also good, except at angles smaller than 20° where the approximation of small deviations is no longer valid.

Appendix 5: Path Interference Model

Ion-Atom Collisions

A projectile P has a constant velocity $\overrightarrow{v_n}$ during the collision. After double capture, the projectile emits an electron with a velocity $\overrightarrow{k_p}$ (Fig. A.12) that will be detected at an angle θ_d with respect to the incident beam direction. The path interference model assumes that two different paths lead to the same detection angle. The electron can be detected directly (1), or can be detected after it has scattered on the target (2).

Both paths lead to a phase difference due to the delay that occurs during the path (2). In the following, we suppose that the amplitude associated with the electron can be written as

$$\overline{A_o} = A_o e^{-i\frac{E}{\hbar}t} \tag{A.41}$$

At any point M of the detector, far from the collision point, the amplitudes corresponding to paths (1) and (2) are

$$\begin{aligned} \overline{A_1} &= A_o e^{-i\frac{E}{\hbar}\left(t - \frac{PM}{k_p}\right)} \\ \overline{A_2} &= A_o e^{-i\frac{E}{\hbar}\left(t - \frac{PT+TM}{k_p}\right)} \end{aligned} \tag{A.42}$$

Because the two paths are indistinguishable, the amplitudes in Eq. A.42 have to be added. The total amplitude is thus $\overline{A}(M) = \overline{A_1} + \overline{A_2}$, and the total intensity can be written $I(M) = \left|\overline{A_1} + \overline{A_2}\right|^2$

$$I(M) = \left| A_o e^{-i\frac{E}{\hbar}\left(t - \frac{PM}{k_p}\right)} + A_o e^{-i\frac{E}{\hbar}\left(t - \frac{PT+TM}{k_p}\right)} \right|^2$$

F. Frémont, *Young-Type Interferences with Electrons*, Springer Series on Atomic, Optical, and Plasma Physics 77, DOI 10.1007/978-3-642-38479-0,

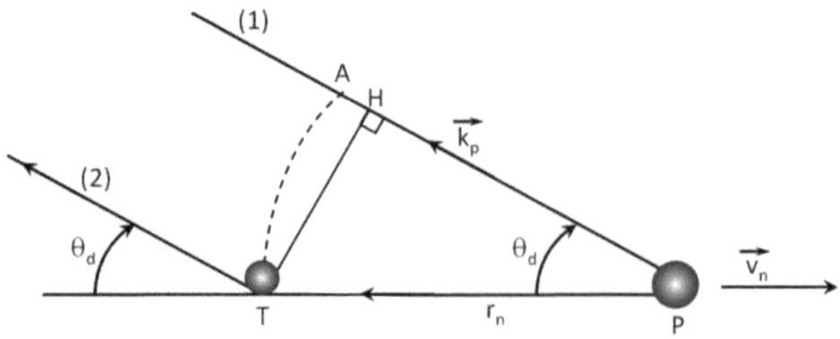

Fig. A.12 Collision between a projectile P and a target T. The electron is emitted with an angle θ_d, and reaches the detector either directly (1) or after scattering on the target (2)

To simplify, we take $t = 0$ when the electron is emitted. The above expression gives

$$I(M) = 2A_o^2(1 + \cos\varphi) \tag{A.43}$$

with $\varphi = \frac{k_p}{2}(PM - PT - TM)$.

$$PM - PT - TM = PM - PA - HM = PA - PH$$

$$= r_n - r_n \cos\theta_d = r_n - \overrightarrow{r_n}.\frac{\overrightarrow{k_p}}{k_p}$$

$$\varphi = \frac{k_p}{2}\left(r_n - \overrightarrow{r_n}.\frac{\overrightarrow{k_p}}{k_p}\right) = \frac{1}{2}\left(k_p r_n - \overrightarrow{r_n}.\overrightarrow{k_p}\right)$$

$$I(M) = 2A_o^2\left(1 + \cos\frac{1}{2}\left(k_p r_n - \overrightarrow{k_p}.\overrightarrow{r_n}\right)\right) \tag{A.44}$$

Ion-Molecule Collisions

The case of ion molecule collisions is described schematically in Fig. A.13. The target consists of two protons labeled T_1 and T_2. The electron that is emitted in the directions of the protons with an angle θ_d can reach the detector either directly (3) or after it has scattered on one or other of the protons (paths (1) and (2)). Therefore, the total emission amplitude is the sum of three combinations, i.e. $\overline{A}(M) = \overline{A_1} + \overline{A_2} + \overline{A_3}$, where $\overline{A_i}$ $(i = 1,2,3)$ is associated to path (i). Here, only the combination of $\overline{A_1}$ and $\overline{A_2}$ is of interest, since they give rise to Young-type interferences.

The intensity *I(M)* associated with paths (1) and (2) is

$$I(M) = 2A_o^2(1 + \cos\psi) \tag{A.45}$$

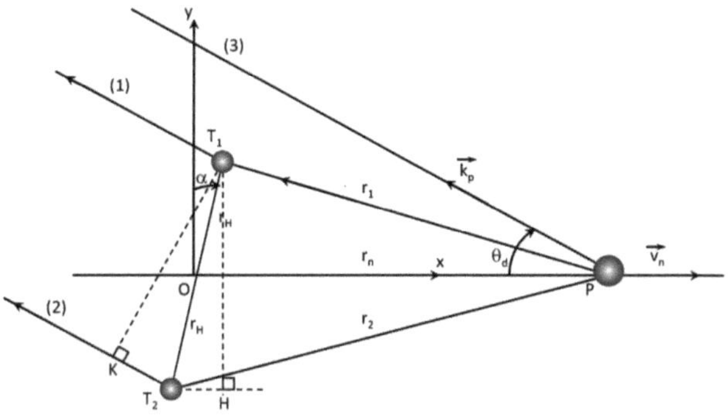

Fig. A.13 Collision between a projectile P and a molecular target composed of two protons T_1 and T_2. The electron is emitted with an angle θ_d, and reaches the detector either directly (3) or after scattering on each proton ((1) and(2))

with $\psi = \frac{k_p}{2}(r_2 - r_1 + T_2M - T_1M)$.

At large distances r_1 and r_2, the difference $r_2 - r_1$ is of the order of T_2H. The difference $T_2M - T_1M$ is equal to T_2K.

$$T_2H = d_{HH}\cos\alpha = \overrightarrow{d_{HH}} \cdot \frac{\overrightarrow{v_n}}{v_n}$$

$$T_2H = \overrightarrow{d_{HH}} \cdot \frac{\overrightarrow{k_p}}{k_p}$$

$$\psi = \frac{k_p}{2}\left(\overrightarrow{d_{HH}} \cdot \frac{\overrightarrow{v_n}}{v_n} + \overrightarrow{d_{HH}} \cdot \frac{\overrightarrow{k_p}}{k_p}\right) = \frac{k_p}{2}\overrightarrow{d_{HH}} \cdot \left(\frac{\overrightarrow{v_n}}{v_n} + \frac{\overrightarrow{k_p}}{k_p}\right)$$

$$I(M) = 2A_o^2\left(1 + \cos\frac{k_p}{2}\overrightarrow{d_{HH}} \cdot \overrightarrow{s}\right) \tag{A.46}$$

with $\overrightarrow{s} = \frac{\overrightarrow{v_n}}{v_n} + \frac{\overrightarrow{k_p}}{k_p}$.

The expression (A.46) is valid for a fixed angle α between the molecule axis and the beam direction. In the present experiment, all of the molecular orientations are possible. Therefore, the intensity has to be integrated in all of the orientations, thus on of all of the angles ξ between $\overrightarrow{d_{HH}}$ and $\overrightarrow{s}$. The interference term is

$$\int_0^\pi \left(e^{i\frac{k_p}{2}\overrightarrow{d_{HH}} \cdot \overrightarrow{s}} + e^{-i\frac{k_p}{2}\overrightarrow{d_{HH}} \cdot \overrightarrow{s}}\right)\sin\xi\, d\xi = \frac{\sin\delta}{\delta}$$

with $\delta = d_{HH}\left|\overrightarrow{k_p} + k_p\frac{\overrightarrow{v_n}}{v_n}\right| = d_{HH}\left(k_p^2 + k_p^2 + 2k_p^2\cos\theta_d\right)^{1/2} = 2\,d_{HH}\,k_p\cos\theta_d/2$

The interferences are well described by what is referred to as the Debye-Ehrenfest term, which is a Bessel function of order 0

$$\frac{\sin\delta}{\delta} \tag{A.47}$$

where

$$\delta = 2\, d_{HH}\, k_p \cos\theta_d/2 \tag{A.48}$$

Parameters Involved in the Model

The interference pattern depends, according to relation (A.48), on the distance d_{HH} between the protons and the velocity k_p of the electron. Intuitively, these quantities can be defined in a simple way. The quantities d_{HH} and k_p can be defined as the distance when the electron is emitted and the velocity of the electron in the laboratory frame, respectively. Nevertheless, d_{HH} and k_p can also be defined as the distance when the electron scatters on the protons and the velocity of the electron in the laboratory (target) frame, respectively.

To be more specific, d^e_{HH} and k_t are the quantities that describe the most intuitive case, whereas d^s_{HH} and k_p are used in the second case we described. Therefore, four situations can be considered, depending on the couple of values used. Depending on each couple of values for the distance and the velocity, the interference term gives rise to oscillations that do not have the same behavior when interference parameters, such as the wavelength, change. The importance of the choice of model parameters is discussed below in the case of $He^{2+} + H_2$ collisions.

Applications to Young-Type Double Slit Interferences

As explained above, four couples of values for the distance d_{HH} and velocity k can be chosen. Figure A.14 shows for example the evolution of the oscillation period calculated using these different values. The evolution is compared with the experimental period T_{exp} (open circles) found for the collision system $He^{2+} + H_2$ at projectile velocities v_p of 0.63 and 0.32 a.u. The period T_{exp} is found to decrease slowly with decreasing v_p. This trend is coherent with the full curve, which corresponds to the distance d^e_{HH} and the velocity k_t, which are the intuitive values described in Sect. 5.3.

The model has also been applied to the angular distribution at both velocities, and compared with the experimental angular distributions. The experiment revealed that the angles of the oscillations shifted, so that maxima (or minima) were replaced by

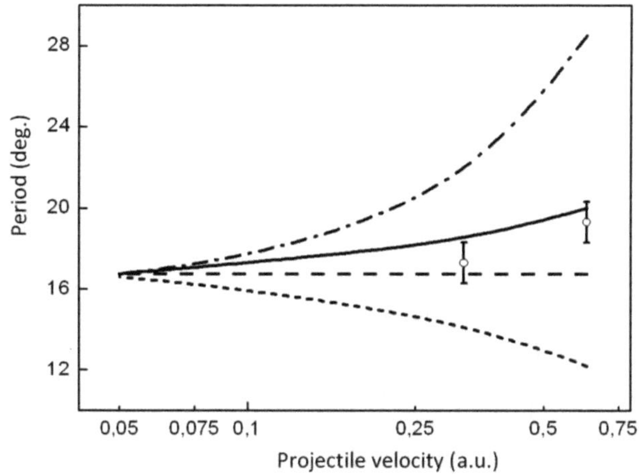

Fig. A.14 Period of the oscillations as a function of projectile velocity. Calculations: (d^e_{HH},k_p), *dotted-dashed curve*; (d^s_{HH},k_p), *dashed curve*; (d^e_{HH},k_t), *full curve*; (d^s_{HH}, k_t), *dotted curve*. The experimental periods are represented by *open circles*

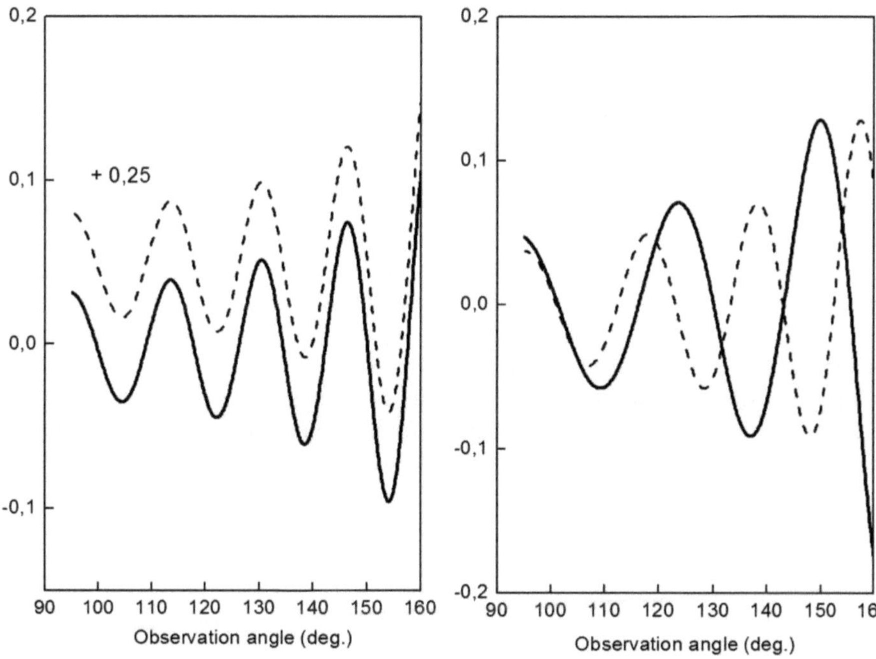

Fig. A.15 Amplitude of the oscillations calculated with the free parameters (d^e_{HH}, k_p) (left-hand side) and (d^s_{HH}, k_p) (right-hand side) values, as a function of the observation angle. *Dashed curve*, 8 keV; *full curve*, 30 keV

minima (or maxima). When k_p was used (Fig. A.15), the calculations did not reproduce the experiment, whatever the distance involved. In the particular case of d^e_{HH}, the phase shift was only similar to that of the experiment for angles greater than 140°.

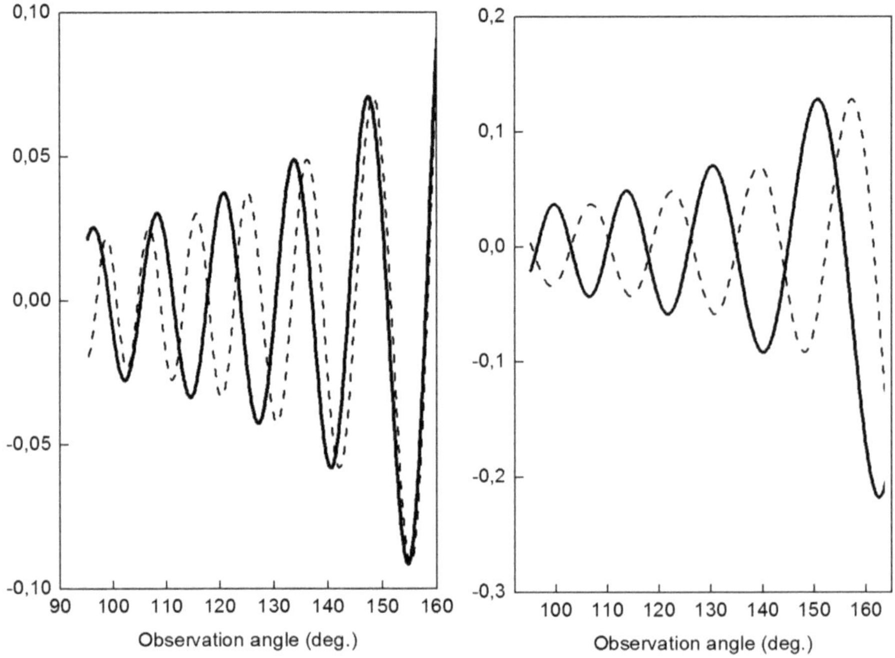

Fig. A.16 Amplitude of the oscillations (d^e_{HH}, k_t) (left-hand side) and (d^s_{HH}, k_t) (right-hand side) values. *Dashed curve*, 8 keV; *full curve*, 30 keV

When k_t is used (Fig. A.16) instead of k_p, the best agreement is observed for the most intuitive value of d_{HH} (right-hand side of Fig. A.16), confirming that the interferences already occur when the electron is emitted, in contrast with what happens for photons, which interfere when they pass through the interferometer.

If d^e_{HH} is used (left-hand side of Fig. A.16), no phase shift is observed. Rather, the oscillation period is changed (see also Fig. A.14).

In conclusion, the path-interference model is a powerful tool for describing the interference characteristics, and predicting, for a given collision system, the period of the oscillations and their behavior when the interference parameters are changed.

Appendix 6: Oscillations in the Width and Wavelength Distribution in Photon Interference Experiments

Simplification of the Problem

To demonstrate the possibility of oscillations in the width and wavelength distributions in the case of photon interference experiments, and to simplify the calculations, the diffraction pattern produced by the edge of the mirror (Lloyd mirror interferences) or by the slits (Young double-slit interferences) is not taken into account. Therefore, we suppose an initial Gaussian profile centered on λ_o with a FWHM $\Delta\lambda_o$ (top of Fig. A.17).

After the light has passed through the interferometer, it is supposed that the profile is disturbed by an interference term. Therefore, the intensity at fixed point M is

$$I(x,\lambda) = e^{-4\ln 2\left(\frac{\lambda-\lambda_o}{\Delta\lambda_o}\right)^2}\cos^2\frac{ax}{\lambda} \tag{A.49}$$

where, in the interference term, $a = \pi d/D$ (d is the distance between the two virtual sources and D the distance between the source center and the screen). The resulting profile can be approximated with a Gaussian centered at $\lambda_{max} \sim \lambda_o$ with a width $\Delta\lambda \sim \Delta\lambda_o$. With expression (A.49), both λ_{max} and $\Delta\lambda$ quantities can be determined analytically.

Determination of the Width Distribution

In principle, the final profile is no longer a Gaussian curve. However, we suppose that the profile is symmetrical around λ_{max}. The width $\Delta\lambda$ is defined from the relation

$$\frac{1}{2}I(\lambda_{\max}) = I\left(\lambda_{\max} \pm \frac{\Delta\lambda}{2}\right) \tag{A.50}$$

F. Frémont, *Young-Type Interferences with Electrons*, Springer Series on Atomic, Optical, and Plasma Physics 77, DOI 10.1007/978-3-642-38479-0,

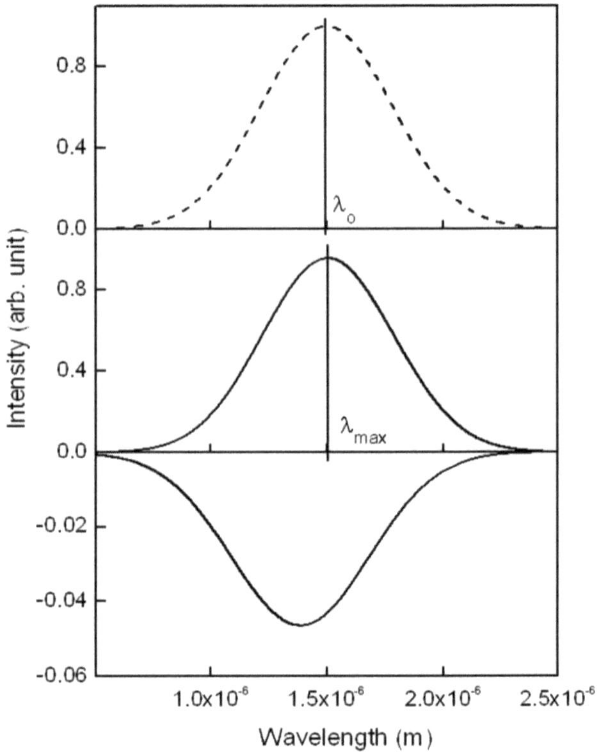

Fig. A.17 Initial Gaussian profile for the wavelength distribution at a fixed position on the detector. After passing through the slits, the profile is distorted (*middle*). The difference between these two profiles is shown in the lower part

$$\frac{1}{2}\cos^2\frac{ax}{\lambda_{\max}} = e^{-4\ln 2\left(\frac{\lambda_{\max}\pm\Delta\lambda/2-\lambda_{\max}}{\Delta\lambda}\right)^2}\cos^2\frac{ax}{\lambda_{\max}\pm\Delta\lambda/2}$$

Assuming symmetry on either side of λ_{max}, the calculation is performed only on the right side. Moreover, we maintain the approximation $\lambda_{max} \sim \lambda_o$, since the variations in the width are expected to be greater than the variations in the wavelength.

Determination of the Wavelength Distribution

We can write $\Delta\lambda = \Delta\lambda_o + \delta\lambda$, so that

$$I\left(\lambda_{\max}+\frac{\Delta\lambda}{2}\right) \approx I\left(\lambda_{\max}+\frac{\Delta\lambda_o}{2}\right) + \delta\lambda\frac{dI}{d\lambda}\bigg|_{\lambda_{\max}+\frac{\Delta\lambda_o}{2}}.$$

Expression (A.50) thus becomes

$$\cos^2\frac{ax}{\lambda_o} = \cos^2\frac{ax}{\lambda_o + \frac{\Delta\lambda_o}{2}} + \delta\lambda\left[-\frac{4\ln 2}{\lambda_o}\cos^2\frac{ax}{\lambda_o + \frac{\Delta\lambda_o}{2}} + \frac{ax}{\lambda_o + \frac{\Delta\lambda_o}{2}}\sin\frac{2ax}{\lambda_o + \frac{\Delta\lambda_o}{2}}\right]$$

$$\delta\lambda = \frac{\cos^2\frac{ax}{\lambda_o} - \cos^2\frac{ax}{\lambda_o + \Delta\lambda_o/2}}{-\frac{4\ln 2}{\lambda_o}\cos^2\frac{ax}{\lambda_o + \Delta\lambda_o/2} + \frac{ax}{\lambda_o + \Delta\lambda_o/2}\sin\frac{2ax}{\lambda_o + \Delta\lambda_o/2}} \tag{A.51}$$

The above expression can be simplified, remembering that $\lambda_o >> \Delta\lambda_o/2$. Consequently, a Taylor development of the different functions involved in relation (A.51) cans be easily performed.

$$\cos^2\frac{ax}{\lambda_o + \Delta\lambda_o/2} = \cos^2\frac{ax}{\lambda_o\left(1 + \frac{\Delta\lambda_o/2}{\lambda_o}\right)} \approx \cos^2\left(\frac{ax}{\lambda_o} - \frac{ax\Delta\lambda_o}{2\lambda_o^2}\right)$$

$$\approx \left(\cos\frac{ax}{\lambda_o}\cos\frac{ax\Delta\lambda_o}{2\lambda_o^2} + \sin\frac{ax}{\lambda_o}\sin\frac{ax\Delta\lambda_o}{2\lambda_o^2}\right)^2 \approx \cos^2\frac{ax}{\lambda_o} + \frac{ax\Delta\lambda_o}{2\lambda_o^2}\sin\frac{2ax}{\lambda_o}$$

$$\sin\frac{2ax}{\lambda_o + \Delta\lambda_o/2} = \sin\frac{2ax}{\lambda_o\left(1 + \frac{\Delta\lambda_o/2}{\lambda_o}\right)} \approx \sin\left(\frac{2ax}{\lambda_o} - \frac{2ax\Delta\lambda_o}{\lambda_o^2}\right)$$

$$\approx \sin\frac{2ax}{\lambda_o}\cos\frac{2ax\Delta\lambda_o}{\lambda_o^2} - \cos\frac{ax}{\lambda_o}\sin\frac{2ax\Delta\lambda_o}{\lambda_o^2} \approx \sin\frac{2ax}{\lambda_o} - \frac{2ax\Delta\lambda_o}{\lambda_o^2}\cos\frac{2ax}{\lambda_o}$$

Finally, expression (A.51) becomes

$$\delta\lambda = \frac{ax\Delta\lambda_o}{\lambda_o}\frac{\sin\frac{ax}{\lambda_o}}{4\ln 2\cos\frac{ax}{\lambda_o} + \frac{2ax}{\lambda_o}\sin\frac{ax}{\lambda_o}} \tag{A.52}$$

It is easily shown that if $\Delta\lambda_o = 0$ (monochromatic source), the variation of the wavelength is 0. In other words, the variations in the width are due to the fact that the source is not monochromatic. Figure A.18 shows the quantity $\delta\lambda$ as a function of the position of the detector. The result is qualitatively similar to that obtained experimentally. A periodic but non-sinusoidal variation of $\delta\lambda$ is observed, as in the experiment. The period of the oscillations is equal to the period of the intensity (top of Fig. A.18).

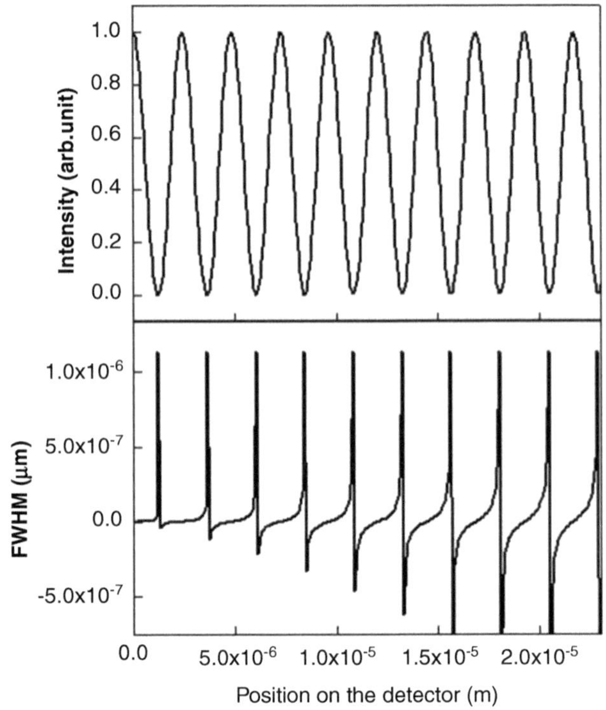

Fig. A.18 Intensity (*top of the figure*) and FWHM oscillations (*bottom*) in photon experiment using a polychromatic source emitting light with a wavelength of 1.532 μm and a FWHM of 0.93 nm

Determination of λ_{max}

The position of the maximum of the energy distribution is obtained when $\frac{dI}{d\lambda} = 0$.

$$-\frac{8\ln 2}{\Delta\lambda_o^2}(\lambda_{\max} - \lambda_o)\cos\frac{ax}{\lambda_{\max}} + \frac{2ax}{\lambda_{\max}^2}\sin\frac{ax}{\lambda_{\max}} = 0$$

We write
$\lambda_{\max} = \lambda_o + \varepsilon$, therefore

$$-\frac{8\ln 2}{\Delta\lambda_o^2}\varepsilon\cos\frac{ax}{\lambda_{\max}} + \frac{2ax}{\lambda_{\max}^2}\sin\frac{ax}{\lambda_{\max}} = 0$$

In the first approximation, we write $\lambda_{max} \sim \lambda_o$ in the above relation. Thus,

$$\lambda_{\max} \approx \lambda_o + \frac{ax}{4\ln 2\lambda_o^2}\Delta\lambda_o^2\tan\frac{ax}{\lambda_o} \tag{A.53}$$

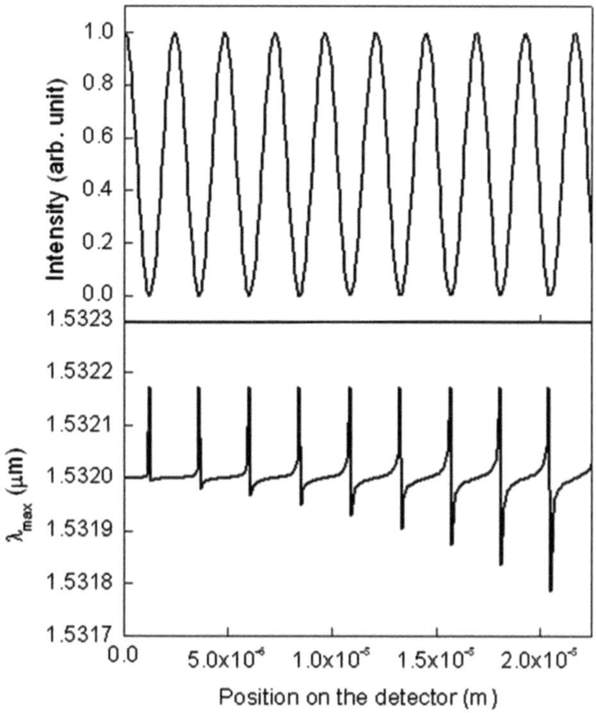

Fig. A.19 Intensity (*top of the figure*) and λ_{max} (*bottom*) in a photon experiment using a polychromatic source emitting light with a wavelength of 1.532 μm and a FWHM of 0.93 nm. The wavelength is calculated using relation A.53

The result of relatiàon (A.53) is represented in Fig. A.19 (bottom) as a function of the position of the detector, and compared with the intensity (top of the figure). As in the calculation of the FWHM, the maximum of the wavelength oscillates around λ_o, with an amplitude of less than 1/1,000. The oscillations, as in the case of the FWHM, are periodic, with the same period as that of the intensity, but not sinusoidal.

Appendix 7: Oscillations at Maximum of Electron Energy Distribution

Experimental and Calculated Energy Maxima

From the experimental fit (cf. Chap. 4 Sect. 4.3), the quantity E_{max} for the $2s^2\ ^1S$ and ($2p^2\ ^1D$ – $2s2p\ ^1P$) configurations were determined. These are represented in Fig. A.20 (open circles) as a function of the detection angle in the 110–160° range.

In addition, the theoretical values for the energy maximum in the laboratory frame E_o are plotted (full curves). These are determined using the relation

$$E_o = \left[\left(\frac{E_p}{M}\right)^{1/2} \cos\theta_d + \left(E_e - \frac{E_p}{M}\sin^2\theta_d\right)^{1/2}\right]^2 \quad \text{(A.54)}$$

where E_p and M are the projectile energy and mass, respectively, E_e is the Auger electron energy in the projectile frame and θ_d is the detection angle.

In the calculation, post-collision effects, which depend on θ_d and the mean lifetime Γ, were not included. This is the main reason for the shift in energy observed for both configurations. To a lesser extent, the other reason for the shift is the uncertainty of the detection angle.

Despite this shift, a detailed analysis of E_{max} angular distribution is possible because, as shown in the next section, the oscillation amplitude is proportional to Γ. Since Γ ($2p^2\ ^1D$) is smaller than Γ ($2s^2\ ^1S$), the oscillations for $2p^2\ ^1D$ can be ignored. Thus, the oscillations for $2s^2\ ^1S$ can be analyzed by matching the experimental and calculated energy for the ($2p^2\ ^1D$, $2s2p\ ^1P$) configurations (Fig. A.21).

To obtain evidence of possible oscillations in E_{max} distribution for the $2s^2\ ^1S$ configuration, the corrected experiment energy was divided by the corresponding calculated energy. The result of the division is presented in Fig. A.22. In the range from 134° to 160°, oscillations are clearly visible, again with the same period as that observed for the intensity. As expected, the amplitude is small, since it represents less than 0.2 % of E_{max}. No oscillation can be noted below 134° since the uncertainties are too great relative to the amplitude.

F. Frémont, *Young-Type Interferences with Electrons*, Springer Series on Atomic, Optical, and Plasma Physics 77, DOI 10.1007/978-3-642-38479-0,

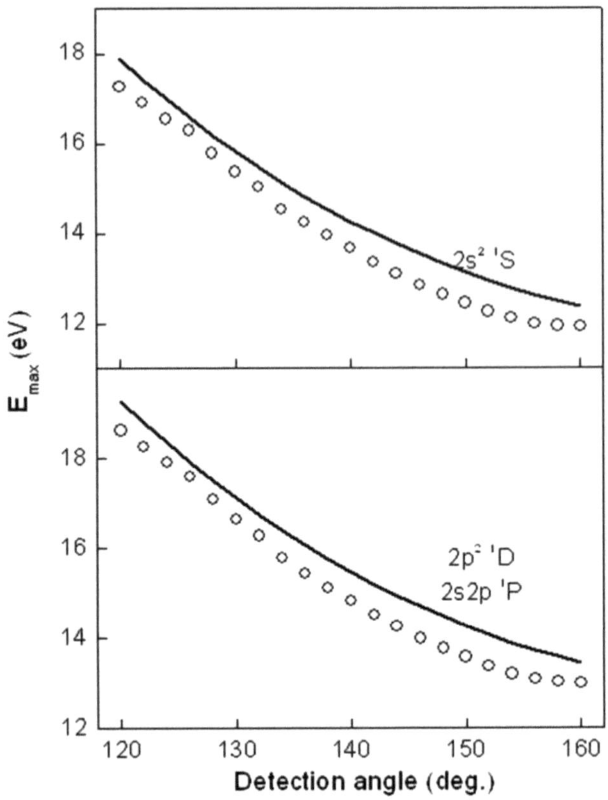

Fig. A.20 Maximum of energy E_{max} for the $2s^2\ ^1S$ (*top*) and ($2p^2\ ^1D$ – $2s2p\ ^1P$) (*bottom*) configurations. *Open circles*, experiment; *full curve*, calculation using relation (A.54)

Determination of Maximum of the Energy Distribution

In the absence of protons, the energy profile can be assimilated with a Gaussian curve. Owing to the vicinity of the protons, the profile is distorted, and oscillations appear due to Young-type interferences. To simplify the calculations, the distortion caused by the post-collision effect is ignored, so that the intensity is of the form

$$I(E) = e^{-\frac{(E-E_O)^2}{\sigma^2}}\left(1 + \frac{\sin(\delta + \varphi)}{\delta}\right) \tag{A.55}$$

where $\delta = 2d\cos\frac{\theta}{2}\left[\left(2E - v_p^2\sin^2\theta\right)^{1/2} + v_p\cos\theta\right]$ in the laboratory frame.

Suppose, to simplify the expression, that the profile is also valid in the projectile frame, and that $\delta = 2d\sqrt{2E}\cos\frac{\theta}{2}$

In this case, $I = e^{-\frac{(E-E_O)^2}{\sigma^2}}\left(1 + \frac{\sin\left(2d\sqrt{2E}\cos\frac{\theta}{2} + \varphi\right)}{2d\sqrt{2E}\cos\frac{\theta}{2}}\right)$

We write $\alpha(\theta) = 2\sqrt{2}d\cos\frac{\theta}{2}$

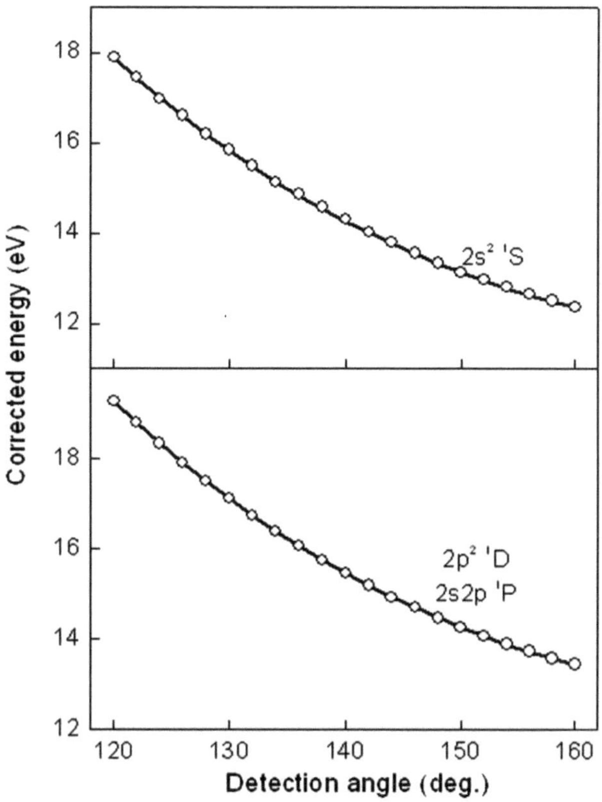

Fig. A.21 Maximum of energy E_{max} for the $2s^2\ ^1S$ (*top*) and ($2p^2\ ^1D$ – $2s2p\ ^1P$) (*bottom*) configurations. *Open circles*, experiment; *full curve*, calculation using relation (A.54). The experiment for ($2p^2\ ^1D$ – $2s2p\ ^1P$) configuration has been obtained after matching it to the calculation

$$I = e^{-\frac{(E-E_O)^2}{\sigma^2}}\left(1+\frac{\sin\left(\alpha(\theta)\sqrt{E}+\varphi\right)}{\alpha(\theta)\sqrt{E}}\right)$$

For a fixed angle, the maximum oscillates since E is inside a sine function. We call E_m the maximum. Since E_m is close to E_o, we can write $E_m = E_o + \delta E$.

$$\frac{dI}{dE}=\frac{de^{-\frac{(E-E_O)^2}{\sigma^2}}}{dE}\left(1+\frac{\sin\left(\alpha(\theta)\sqrt{E}+\varphi\right)}{\alpha(\theta)\sqrt{E}}\right)+e^{-\frac{(E-E_O)^2}{\sigma^2}}\frac{d\sin\left(\alpha(\theta)\sqrt{E}+\varphi\right)}{dE}\frac{1}{\alpha(\theta)\sqrt{E}}$$
$$+e^{-\frac{(E-E_O)^2}{\sigma^2}}\sin\left(\alpha(\theta)\sqrt{E}+\varphi\right)\frac{d\frac{1}{\alpha(\theta)\sqrt{E}}}{dE}$$

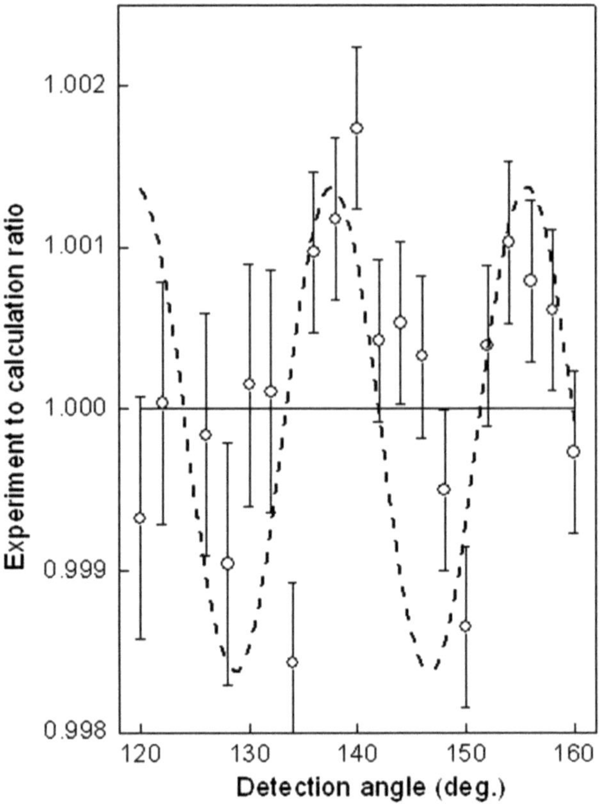

Fig. A.22 Experiment to calculation ratio for the $2s^2$ ^1S configuration (*open circles*). The sinusoidal curve (*dashed line*) is only plotted as a guide for the eye

$$\frac{dI}{dE} = -2\frac{(E - E_o)}{\sigma^2} e^{-\frac{(E-E_o)^2}{\sigma^2}} \left(1 + \frac{\sin\left(\alpha(\theta)\sqrt{E} + \varphi\right)}{\alpha(\theta)\sqrt{E}}\right)$$
$$+ e^{-\frac{(E-E_o)^2}{\sigma^2}} \frac{\alpha(\theta)}{2\sqrt{E}} \cos\left(\alpha(\theta)\sqrt{E} + \varphi\right) \frac{1}{\alpha(\theta)\sqrt{E}}$$
$$+ e^{-\frac{(E-E_o)^2}{\sigma^2}} \sin\left(\alpha(\theta)\sqrt{E} + \varphi\right) \left(-\frac{1}{2}\right) \frac{1}{\alpha(\theta)E\sqrt{E}}$$

$$\frac{dI}{dE} = 0$$

$$0 = -2\frac{\delta E}{\sigma^2}\left(1 + \frac{\sin\left(\alpha(\theta)\sqrt{E_o + \delta E} + \varphi\right)}{\alpha(\theta)\sqrt{E_o + \delta E}}\right)$$
$$+ \frac{\alpha(\theta)}{2\sqrt{E_o + \delta E}}\cos\left(\alpha(\theta)\sqrt{E_o + \delta E} + \varphi\right)\frac{1}{\alpha(\theta)\sqrt{E_o + \delta E}}$$
$$+ \sin\left(\alpha(\theta)\sqrt{E_o + \delta E} + \varphi\right)\left(-\frac{1}{2}\right)\frac{1}{\alpha(\theta)(E_o + \delta E)^{3/2}}$$

Since $E_o >> \delta E$:

$$E_o \sin\left(\alpha(\theta)\sqrt{E_o + \delta E} + \varphi\right) \approx \sin\left(\alpha(\theta)\sqrt{E_o} + \varphi\right) + \frac{\alpha(\theta)\delta E}{2\sqrt{E_o}} \times \cos\left(\alpha(\theta)\sqrt{E_o} + \varphi\right)$$

$$\cos\left(\alpha(\theta)\sqrt{E_o + \delta E} + \varphi\right) \approx \cos\left(\alpha(\theta)\sqrt{E_o} + \varphi\right) - \frac{\alpha(\theta)\delta E}{2\sqrt{E_o}}\sin\left(\alpha(\theta)\sqrt{E_o} + \varphi\right)$$

$$\frac{1}{\alpha(\theta)(E_o + \delta E)^{1/2}} \approx \sqrt{E_o}\frac{1 - \frac{\delta E}{2E_o}}{\alpha(\theta)} = \frac{\sqrt{E_o} - \frac{\delta E}{2\sqrt{E_o}}}{\alpha(\theta)}$$

$$\frac{1}{\alpha(\theta)(E_o + \delta E)^{3/2}} \approx E_o\sqrt{E_o}\frac{1 - \frac{3\delta E}{2E_o}}{\alpha(\theta)} = \frac{E_o\sqrt{E_o} - \frac{\delta E}{2}\sqrt{E_o}}{\alpha(\theta)}$$

Using these developments, the quantity δE follows

$$0 = -2\frac{\delta E}{\sigma^2}\left(\alpha(\theta)\left(\sqrt{E_o} + \frac{\delta E}{2\sqrt{E_o}}\right) + \sin\left(\alpha(\theta)\sqrt{E_o}\right) + \frac{\alpha(\theta)\delta E}{2\sqrt{E_o}}\cos\left(\alpha(\theta)\sqrt{E_o}\right)\right)$$
$$+ \frac{\alpha(\theta)}{2}\left(\sqrt{E_o} - \frac{\delta E}{2\sqrt{E_o}}\right)\left(\cos\left(\alpha(\theta)\sqrt{E_o}\right) - \frac{\alpha(\theta)\delta E}{2\sqrt{E_o}}\sin\left(\alpha(\theta)\sqrt{E_o}\right)\right)$$
$$-\frac{1}{2}\left(\sin\left(\alpha(\theta)\sqrt{E_o}\right) + \frac{\alpha(\theta)\delta E}{2\sqrt{E_o}}\cos\left(\alpha(\theta)\sqrt{E_o}\right)\right)\left(\sqrt{E_o} - \frac{\delta E}{2\sqrt{E_o}}\right)$$

Let us limit development to the first order:

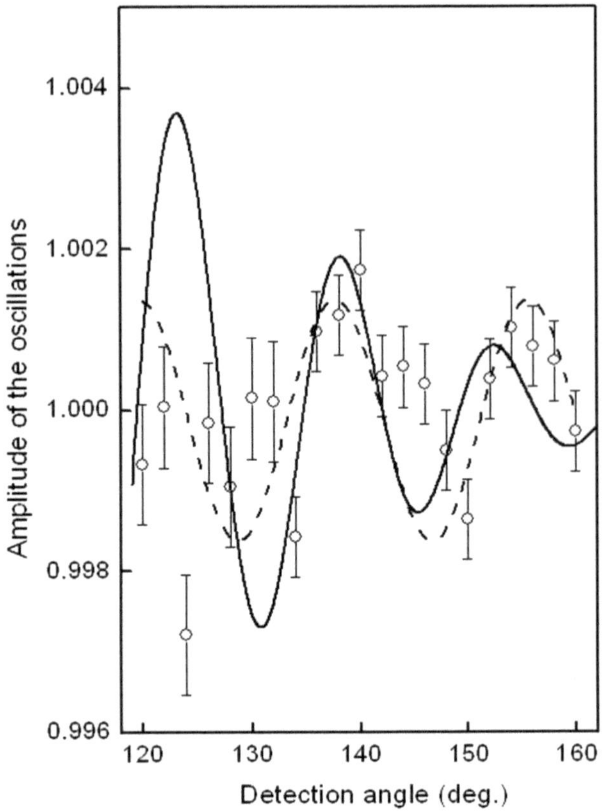

Fig. A.23 Experiment to calculation ratio for the $2s^2$ ^1S configuration (*open circles*). The sinusoidal curve (*dashed line*) is only plotted as a guide for the eye. The *full curve* is determined using the final expression of δE

$$
\begin{aligned}
0 = & -2\frac{\delta E}{\sigma^2}\left(\alpha(\theta)\sqrt{E_o} + \sin\left(\alpha(\theta)\sqrt{E_o}\right)\right) \\
& + \frac{\alpha(\theta)}{2}\left(\sqrt{E_o} - \frac{\delta E}{2\sqrt{E_o}}\right)\left(\cos\left(\alpha(\theta)\sqrt{E_o}\right)\right) \\
& + \frac{\alpha(\theta)}{2}\left(\sqrt{E_o}\right)\left(-\frac{\alpha(\theta)\delta E}{2\sqrt{E_o}}\sin\left(\alpha(\theta)\sqrt{E_o}\right)\right) \\
& - \frac{1}{2}\left(\sin\left(\alpha(\theta)\sqrt{E_o}\right) + \frac{\alpha(\theta)\delta E}{2\sqrt{E_o}}\cos\left(\alpha(\theta)\sqrt{E_o}\right)\right)\left(\sqrt{E_o}\right) \\
& - \frac{1}{2}\left(\sin\left(\alpha(\theta)\sqrt{E_o}\right)\right)\left(-\frac{\delta E}{2\sqrt{E_o}}\right)
\end{aligned}
$$

$$\delta E = \frac{-\frac{\alpha(\theta)}{2}\sqrt{E_o}\cos\left(\alpha(\theta)\sqrt{E_o}\right) + \frac{1}{2}\sin\left(\alpha(\theta)\sqrt{E_o}\right)\sqrt{E_o}}{-\frac{2}{\sigma^2}\left(\alpha(\theta)\sqrt{E_o} + \sin\left(\alpha(\theta)\sqrt{E_o}\right)\right) - \frac{\alpha(\theta)}{2\sqrt{E_o}}\cos\left(\alpha(\theta)\sqrt{E_o}\right) - \frac{\alpha(\theta)^2}{4}\sin\left(\alpha(\theta)\sqrt{E_o}\right) - \frac{1}{4\sqrt{E_o}}\left(\sin\left(\alpha(\theta)\sqrt{E_o}\right)\right)}$$

This expression depends on σ, thus on the lifetime Γ of the structures.

Figure A.23 shows the result of the expression of δE as a function of the detection angle (full curve). Despite the approximations in the initial intensity, the result is in reasonable agreement with the experiment.

Appendix 8: Fast Fourier Transform

Objective of the Method

The Fast Fourier Transform (FFT) is a useful and efficient tool for computing the Discrete Fourier Transform (DFT). It is a method that transforms a function of time into a function of frequency. More generally, when periodic variations of a function exist but are not visible, FFT can reveal them. Suppose, for example, periodic variations in energy superimposed onto a Gaussian function (open circles in Fig. A.24). The oscillations are so small that they are hidden in the main profile, given by a pure Gaussian curve (solid line).

Nevertheless, as shown in Fig. A.25, the FFT of the whole function reveals the presence of oscillations (open circles) by the emergence of a structure located at ~3 eV^{-1}.

Fourier Series Decomposition of a Periodic Function

A function f of variable x is periodic with period T. The function $f(x)$ can be written as an infinite discrete sum of periodic functions

$$f(x) = \sum_{n=-\infty}^{+\infty} c_n \, e^{in2\pi \, \nu \, x} \tag{A.56}$$

where the frequency ν has the dimension of inverse time. The coefficients c_n are calculated from relation (A.56)

$$c_n = \frac{1}{T} \int_{-T/2}^{T/2} f(x) e^{-in2\pi \, \nu \, x} dx \tag{A.57}$$

The function has been decomposed into Fourier series. The coefficients c_n are the Fourier coefficients of the function f. In practice, these coefficients are unusable, since their signals are truncated. Therefore, the definition has to be modified.

F. Frémont, *Young-Type Interferences with Electrons*, Springer Series on Atomic, Optical, and Plasma Physics 77, DOI 10.1007/978-3-642-38479-0,

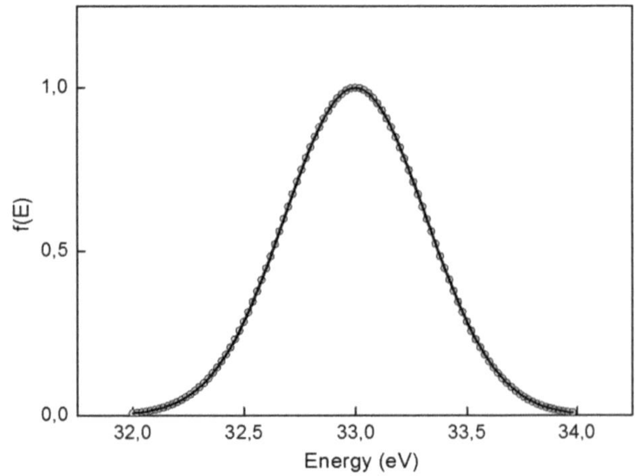

Fig. A.24 *Open circles*: Gaussian curve with oscillating term of small amplitude. *Solid line*: Pure Gaussian curve that fits the previous function

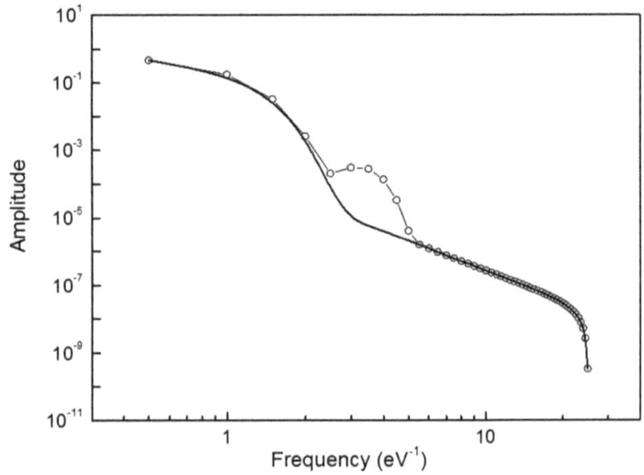

Fig. A.25 *Open circles*: FFT of the Gaussian curve including oscillating term. *Solid line*: FFT of the pure Gaussian curve. The structure located at ~3 eV^{-1} reveals the presence of the oscillations

Fourier Transform of a Non-periodic Function

Definition and Expression

Suppose a continuous function f defined on an interval Δx. The function $f(x)$ can be considered as a periodic function whose period T is infinite. The Fourier transform of f is a function F of variable u defined by

$$F(u) = \frac{1}{\sqrt{2\pi}} \int_{-\infty}^{\infty} f(x) e^{i2\pi\, u\, x} dx \tag{A.58}$$

The quantity u is the conjugate variable of variable x. If x has a time dimension, u is a frequency. If x is an energy, the dimension of u is the inverse of energy. By analogy with time, u will also be called frequency, whatever the dimension of the initial variable x.

Some Examples of Fourier Transforms

Dirac Delta Distribution

The Dirac delta distribution $\delta(x - x_m)$ is zero everywhere except for $x = x_m$. The Fourier transform is

$$F(u) = \frac{1}{\sqrt{2\pi}} \int_{-\infty}^{\infty} \delta(x - x_m)\, e^{i2\pi\, u\, x} dx$$

The real part of this function is $\frac{1}{\sqrt{2\pi}} \cos 2\pi x_m$ with a period of $1/x_m$.

Sine Function

The Fourier transform of a sine function amounts to calculating the inverse Fourier transform of the previous function

$$\frac{1}{\sqrt{2\pi}} \int_{-\infty}^{\infty} \sin ax\, e^{i2\pi\, u\, x} dx = i\sqrt{\frac{\pi}{2}}\delta(u - a) + i\sqrt{\frac{\pi}{2}}\delta(u + a)$$

Rectangular Distribution

The rectangular distribution is defined by *f(x)* = 1 between a and b and is zero elsewhere (left-hand side of Fig. A.26). As an example, let's take $a = -1$ and $b = 1$. In this case, the Fourier transform of *f(x)* is $F(u) = \sin 2\pi u/(2\pi u)$, which is the well known Bessel function of zero order (right-hand side of Fig. A.26).

Fourier Transform of a Sampled Function

Up to now, only the case of continuous functions has been considered. In an experiment, the spectra to be analyzed are not continuous, but sampled. For example, an energy spectrum has a limited number of points. The procedure used above is no longer valid.

Definition and Expression

The discrete Fourier transform of an N term sequence $x(0)$, $x(1)$, … , $x(N - 1)$, is the sequence of N terms $X(0)$, $X(1)$, … , $X(N - 1)$ defined by

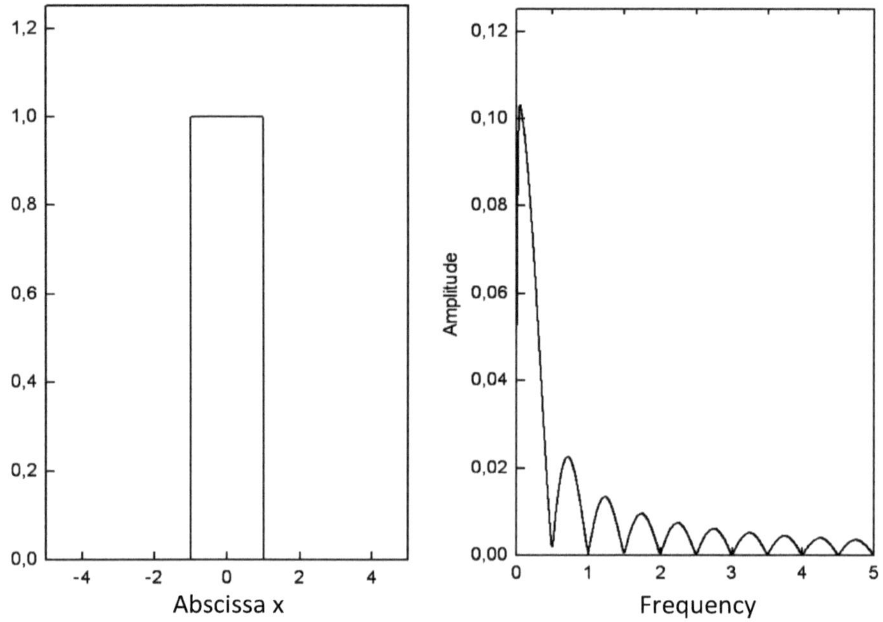

Fig. A.26 Rectangular distribution (*left-hand side*), and absolute value of its Fourier transform (*right-hand side*)

$$X(k) = \sum_{n=0}^{N-1} x(n)\, e^{-i2\pi\frac{n}{N}k} \tag{A.59}$$

In our study, the N values of n are the energies of the emitted electrons. Let us compare the results for both a continuous spectrum and a discreet one. In Fig. A.27 an exponentially decreasing curve is plotted (left-hand side), as well as the amplitude of its Fourier transform.

Suppose that the incoming signal is no longer continuous, but composed of 15 points $x(n)$ such as $x(n) = \exp(-an)$, where a is a constant, and n is separated by half a unit (Fig. A.28).

The Fourier transform of the incoming signal shows, in addition to the Fourier transform of a continuous signal, two replicas of this transform, separated by $1/T$, where T is the periodicity (0.5) of the points. The three structures partially overlap, as seen in Fig. A.28. To avoid this overlap, the continuous sequence is truncated by multiplying the incoming function by a function F whose duration is NT. For example, by adding a rectangular distribution, the effect on the Fourier transform is to separate the three structures, and to show undulations on each peak, which are the signature of the Fourier transform of the rectangular distribution (Fig. A.29).

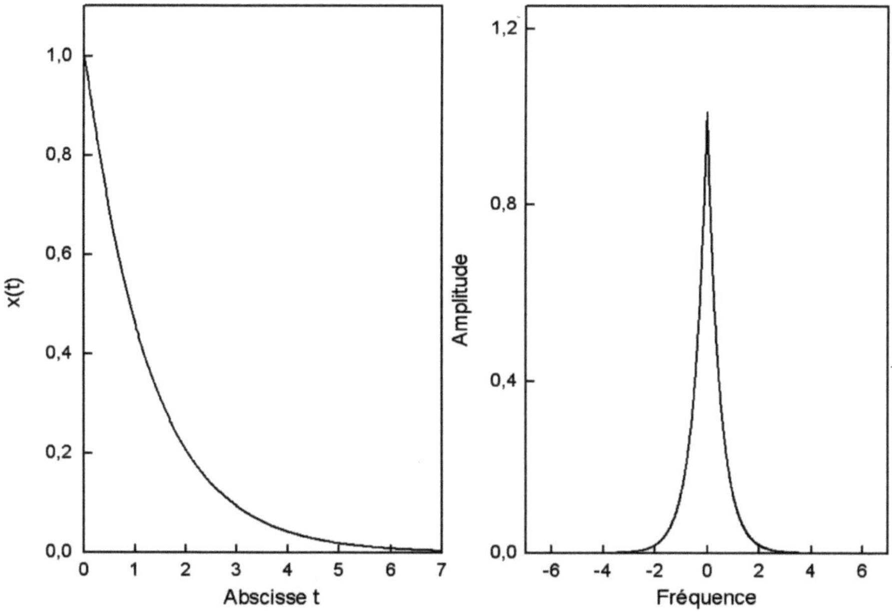

Fig. A.27 Continuous function (*left-hand side*), and amplitude of its Fourier transform (*right-hand side*)

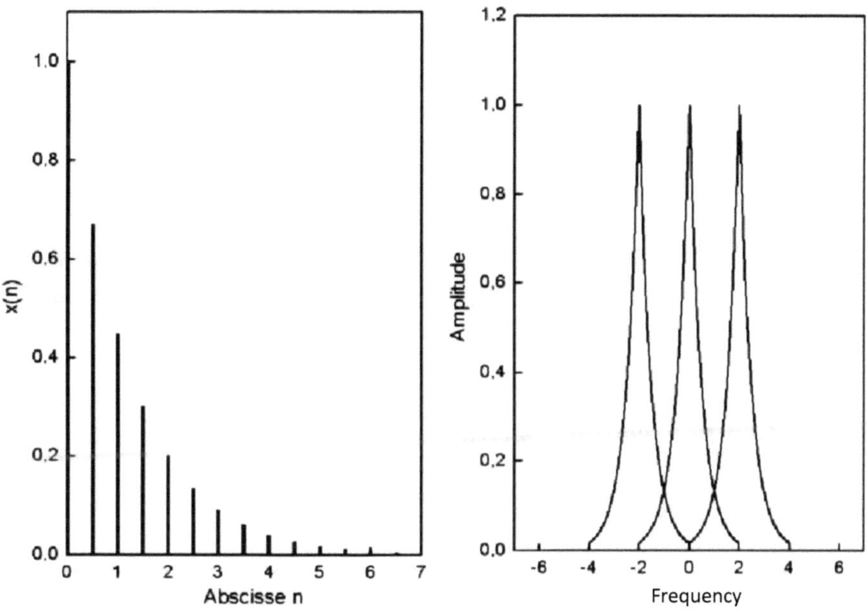

Fig. A.28 Discrete exponentially decreasing function (*left-hand side*), and amplitude of its Fourier transform (*right-hand side*)

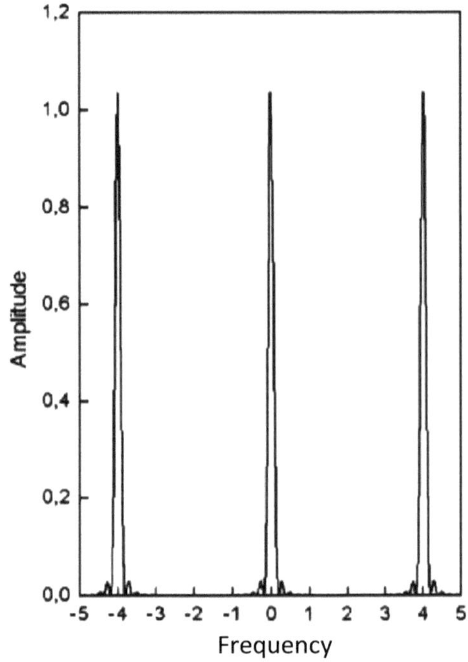

Fig. A.29 Fourier transform of the discrete exponential multiplied by a rectangular function

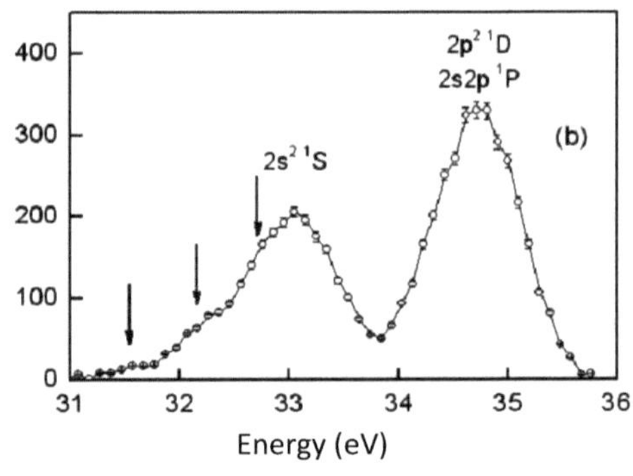

Fig. A.30 Auger spectrum following $2\ell 2\ell'$ configuration deexcitation after doubble electron capture in $He^{2+} + H_2$ collisions

Application to Electron Spectra

Example of a Typical Spectrum

A typical Auger spectrum originating from the capture of two electrons in $He^{2+} + H_2$ collisions is presented in Fig. A.30, after subtraction of the background caused by direct ionization. At a detection angle of 160°, oscillations (arrows point out their maxima) with very weak amplitude are observed, superimposed onto the

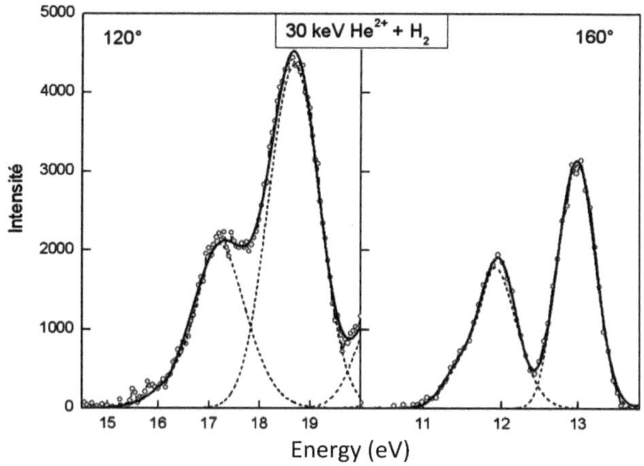

Fig. A.31 Auger spectra at 120° and 160°. At 160°, oscillations are visible between 11 eV and 12.5 eV, while they are no longer visible at 120°. The *dashed* and *full curves* are the result of a fitting procedure using Gaussian curves

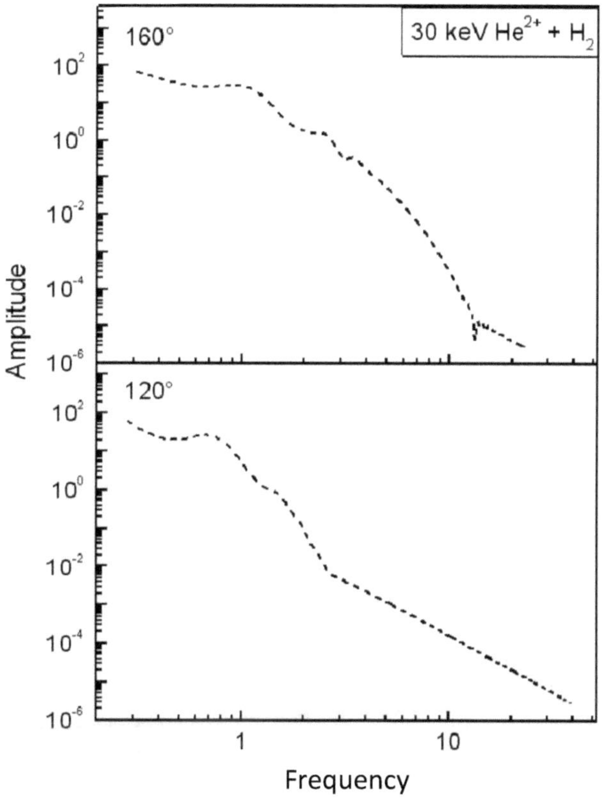

Fig. A.32 Fourier transform of two Gaussian curves

peak associated with the configuration $2s^2\ ^1S$. Analysis by Fourier transform is a very efficient method for obtaining clear evidence of these oscillations.

As mentioned before, the width of the peaks increases as the detection angle decreases. The consequence of this is that the possible existing oscillations are no longer visible, as shown for example in Fig. A.31 for a detection angle of 120°.

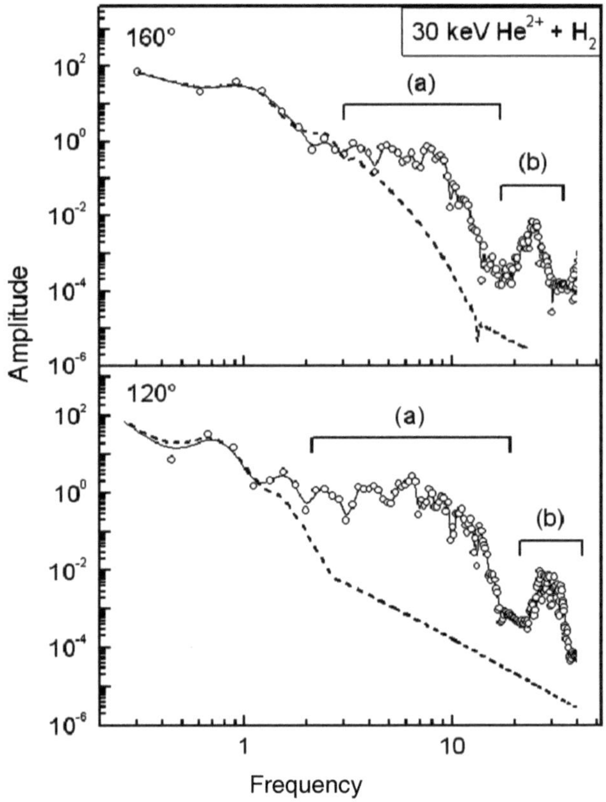

Fig. A.33 Fourier transform of experimental spectra (*open circles*). Fourier transform of the Gaussian curves is also illustrated (*dashed curves*)

To obtain clear evidence of oscillations, the spectra were fitted with Gaussian curves and an improved Fourier transform of these curves was compares them with the Fourier transform of the experimental spectra.

Fourier Transform of Gaussian Curves

When two Gaussian curves are present in a spectrum (Fig. A.31), their Fourier transform is a decreasing function (Fig. A.32)

Two structures (Fig. A.32) appear in the 0.7–3 eV^{-1} range, due to the shift in energy of the two Gaussian curves. The Fourier transform of the experiment (Fig. A.33) shows two additional structures. Structure (b), centered at 30 eV^{-1} is due to statistical fluctuations of the spectra, inducing a period of about 0.03 eV. The large structure (a) is caused by the oscillations that are present in the spectra. Although invisible on the spectra at 120°, these oscillations exist and are revealed by the present method.

Appendix 9: Recoil Ion Momentum and Capture Process

Collision Diagram

Before the collision, the projectile has a mass M, a velocity $\overrightarrow{v_o}$ and a momentum $\overrightarrow{p_o}$ (Fig. A.34). After the collision, the projectile has captured one electron of mass m from a target with the same mass M as the projectile. Its velocity is now $\overrightarrow{v_f}$ and its momentum $\overrightarrow{p_f}$, and its trajectory is deflected with an angle θ_d. On the other side, the target, which is at rest before the collision, recoils with a velocity $\overrightarrow{v_r}$ and a recoil momentum $\overrightarrow{p_r}$. The recoil angle is labeled ψ_r.

The goal of this work is to show the link between the two components $\overrightarrow{p_{//}}$ and $\overrightarrow{p_\perp}$ of $\overrightarrow{p_r}$, and the capture parameters θ_d, ψ_r and Q, defined as the projectile kinetic energy difference before and after the collision.

Energy and Momentum Conservation Laws

Before and after the collision, the total energy and momentum are conserved. These two laws lead, after projection on the axes, to three equations

$$\begin{cases} \frac{p_o^2}{2M} = \frac{p_f^2}{2(M+m)} + \frac{p_r^2}{2(M+m)} + Q & (1) \\ \\ p_o = p_{//} + p_f \cos\theta_d & (2) \\ 0 = p_\perp - p_f \sin\theta_d & (3) \end{cases} \tag{A.60}$$

where Q, also called Q-value, is the difference between the final and initial potential energies.

F. Frémont, *Young-Type Interferences with Electrons*, Springer Series on Atomic, Optical, and Plasma Physics 77, DOI 10.1007/978-3-642-38479-0,

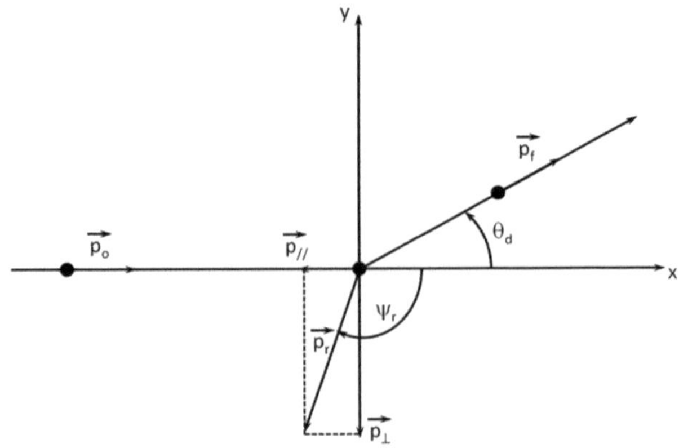

Fig. A.34 Schematic view of the collision. The projectile collides with a target which is at rest before the collision. The target recoils with a momentum $\overrightarrow{p_r}$ whose projections on the x and y axis are $\overrightarrow{p_{//}}$ and $\overrightarrow{p_\perp}$, respectively

Parallel Component of the Recoil Momentum

Equations 2 and 3 are used to extract p_f and $p_\perp$ and to insert the results in Eq. 1

$$(2) : p_f = \frac{p_o - p_{//}}{\cos\theta_d}$$

$$(3) : p_\perp = p_f \sin\theta_d = \left(p_o - p_{//}\right)\tan\theta_d$$

$$\frac{p_o^2}{2M} = \frac{\left(p_o - p_{//}\right)^2}{2(M+m)\cos^2\theta_d} + \frac{p_{//}^2 + p_\perp^2}{2(M+m)} + Q$$

$$\frac{p_o^2}{2M} = \frac{\left(p_o - p_{//}\right)^2}{2(M+m)\cos^2\theta_d} + \frac{p_{//}^2}{2(M+m)} + \frac{\left(p_o - p_{//}\right)^2\tan^2\theta_d}{2(M+m)} + Q$$

$$\frac{2p_{//}^2}{(M+m)\cos^2\theta_d} - \frac{2p_o p_{//}}{(M+m)}\left[\frac{1}{\cos^2\theta_d} + \tan^2\theta_d\right] + p_o^2\left[\frac{1+\sin^2\theta_d}{(M+m)\cos^2\theta_d} - \frac{1}{M}\right] + 2Q$$
$$= 0$$

At projectile velocities of the order of 0.5 a.u., the capture process gives rise to scattering angles θ_d much smaller than 1, so that

$$\cos\theta_d \approx 1$$
$$\sin\theta_d \approx 0$$

The last equation becomes

$$\frac{2p_{//}^2}{(M+m)} - \frac{2p_o p_{//}}{(M+m)} + p_o^2\left[\frac{1}{(M+m)} - \frac{1}{M}\right] + 2Q = 0 \qquad \text{(A.61)}$$

The discriminant of Eq. A.61 is

$$\Delta = \frac{4p_o^2}{(M+m)^2} - 4\frac{2}{(M+m)}\left(p_o^2\left[\frac{1}{(M+m)} - \frac{1}{M}\right] + 2Q\right)$$

$$= \frac{4p_o^2}{(M+m)^2} + 4\frac{2p_o^2}{(M+m)^2}\left(\frac{m}{M} - 2Q\frac{M+m}{p_o^2}\right)$$

$$p_{//} = \frac{M+m}{4}\left(\frac{2p_o}{(M+m)} \pm \left[\frac{4p_o^2}{(M+m)^2} + 4\frac{2p_o^2}{(M+m)^2}\left(\frac{m}{M} - 2Q\frac{M+m}{p_o^2}\right)\right]^{1/2}\right)$$

$$p_{//} = \frac{M+m}{4}\frac{2p_o}{(M+m)}\left(1 \pm \left[1 + 2\left(\frac{m}{M} - 2Q\frac{M+m}{p_o^2}\right)\right]^{1/2}\right)$$

$$p_{//} = \frac{p_o}{2}\left(1 \pm \left[1 + 2\frac{m}{M} - 4Q\frac{M+m}{p_o^2}\right]^{1/2}\right) \tag{A.62}$$

The coefficients m/M and $Q\ (M + m)/p_o^2$ are much smaller than 1. A Taylor expansion of $p_{//}$ can be made

$$p_{//} \approx \frac{p_o}{2}\left(1 \pm \left[1 + \frac{m}{M} - 2Q\frac{M+m}{p_o^2}\right]\right) \approx -p_o\frac{m}{2M} + Q\frac{M+m}{p_o}$$

keeping only the negative solution

$$p_{//} \approx -\frac{p_o}{2M} + \frac{QM}{p_o} \tag{A.63}$$

The present result clearly indicates the link between the determination of $p_{//}$ and the Q-value, which gives insights into the capture process.

Perpendicular Component of the Recoil Momentum

$$0 = p_\perp - p_f \sin\theta_d$$

$$p_\perp = p_f \sin\theta_d \approx p_o\theta_d$$

$$p_\perp \approx p_o\theta_d \tag{A.64}$$

The scattering angle is obtained through the experimental determination of $p_\perp \approx p_o$ θ_d.

Finally, as shown in Chap. 5, the representation of the results in the momentum space needs to be carried out since, as shown in relations (A.63) and (A.64), it is directly connected to the capture parameters, i.e., to the primary process itself.

Appendix 10: Gamma and Hypergeometric Functions

Gamma Function

The Gamma function Γ extends the factorial function to complex numbers. In its integral form, the Gamma function is written

$$\Gamma(z) = \int_0^{\infty} t^{z-1}\, e^{-t}\, dt \tag{A.65}$$

This integral converges when $\mathrm{Re}(z)$ is positive. When $\mathrm{Re}(z)$ is negative, $\Gamma(z)$ is not defined if $\mathrm{Re}(z)$ is an integer (Fig. A.35). When z is a positive integer,

$$\Gamma(z) = (z-1)\,! \tag{A.66}$$

In practice, for better efficiency in the calculation, the form given in Eq. A.65 is not used. The Gamma function is calculated in several ways, depending on the values of z. When $|z|$ is much greater than 1, Stirling's approximation is used

$$\Gamma(z) = e^{-z} z^{z-1/2} (2\,\pi)^{1/2} \sum_{m=0}^{n} \frac{a_m}{z^m} \tag{A.67}$$

If $|z|$ is less than 1, the function $1/\Gamma(z)$ is developed

$$\frac{1}{\Gamma(z)} = \sum_{m=0}^{n} b_m z^m \tag{A.68}$$

F. Frémont, *Young-Type Interferences with Electrons*, Springer Series on Atomic, Optical, and Plasma Physics 77, DOI 10.1007/978-3-642-38479-0,

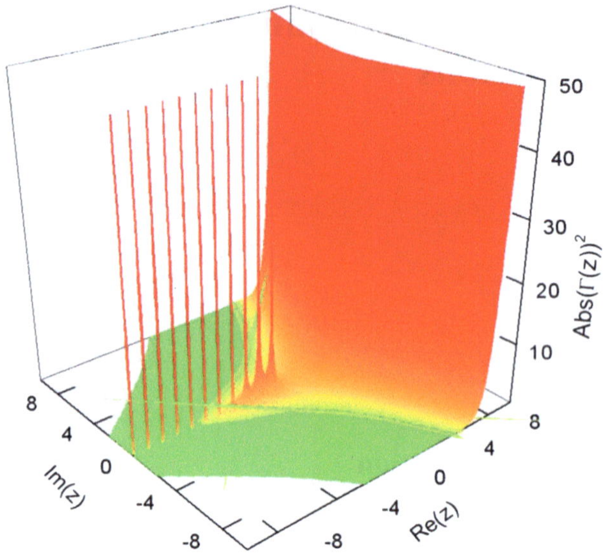

Fig. A.35 Square modulus of $\Gamma(z)$ as a function of real and imaginary parts of z

In expressions (A.67) and (A.68), the index n, which characterizes the last term in the sums, depends on the required precision. The coefficients a_m and b_m are tabulated[5].

Hypergeometric Functions

Definition

The generalized hypergeometric functions are defined as follows

$$ {}_pF_q(a_1, \ldots, a_p; b_1, \ldots, b_q; z) = \sum_{m=0}^{\infty} \frac{(a_1)_m \cdots (a_p)_m}{(b_1)_m \cdots (b_q)_m} \frac{z^m}{m!} \tag{A.69} $$

The coefficients $(x)_m$ inside the sum are called Pochhammer coefficients and are defined by

$$ (x)_m = \frac{\Gamma(x+m)}{\Gamma(x)} \tag{A.70} $$

[5] M. Abramowitz and I.A. Stegun, *Handbook of mathematical functions*.

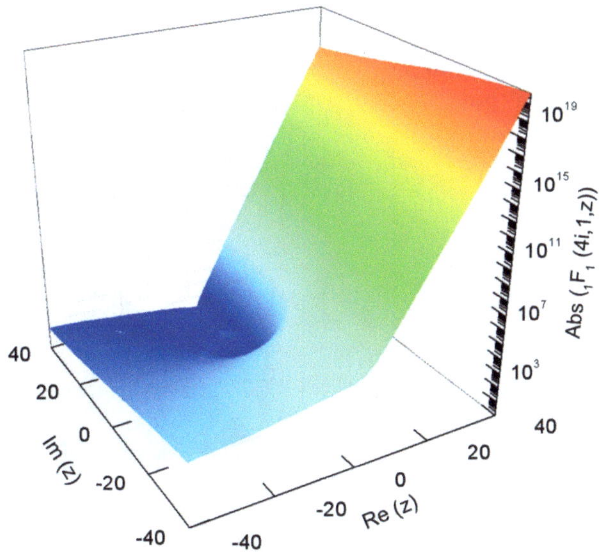

Fig. A.36 Modulus of ${}_1F_1(4i, 1; z)$ as a function of its real and imaginary part

Fonction ${}_1F_1$

Function ${}_1\mathrm{F}_1$ is written

$$
{}_1F_1(a, c; z) = \sum_{m=0}^{\infty} \frac{(a)_m}{(c)_m} \frac{z^m}{m!} \tag{A.71}
$$

and is defined if $|z| < 1$. When $|z|$ is greater than 1, expression (A.71) has to be transformed to insure continuity.

$$
\begin{aligned}
{}_1F_1(a, c; z) \approx & \frac{\Gamma(c)}{\Gamma(c-a)} \frac{1}{z^a} \, {}_2F_o\left(a, a-c+1; -\frac{1}{z}\right) \\
& + \frac{\Gamma(c)}{\Gamma(a)} \frac{e^z}{z^{a-c}} \, {}_2F_o\left(c-a, 1-a; \frac{1}{z}\right)
\end{aligned} \tag{A.72}
$$

The modulus of ${}_1F_1(a, c; z)$ is represented in Fig. A.36 for $a = 4\mathrm{i}$ and $c = 1$ that are typical values in the integrals describing autoionization amplitude using CDW approximation (cf. Chap. 5).

The modulus of ${}_1F_1(4i, 1; z)$ is divided into two regions, for a given value of the imaginary part of z. When $\mathrm{Re}(z)$ is negative, the modulus is nearly constant. In contrast, for positive values of $\mathrm{Re}(z)$, the modulus exponentially increases (note the logarithmic scale).

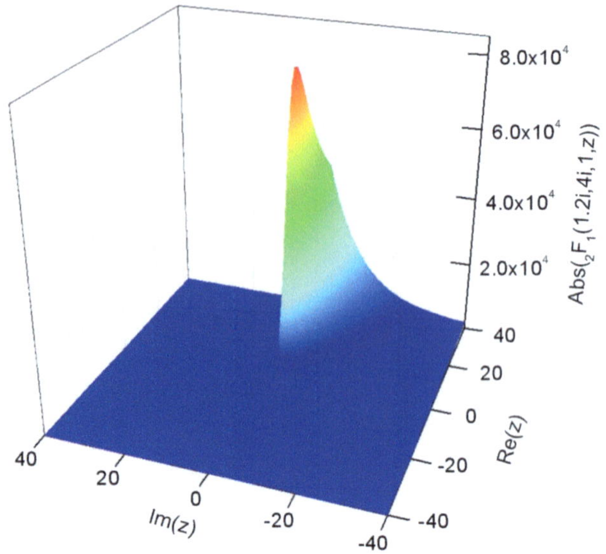

Fig. A.37 Modulus of $_2F_1(1.2i, 1+4i, 1; z)$ as a function of its real and imaginary part

Fonction $_2F_1$

Function $_2F_1$ is written

$$_2F_1(a,b,c;z) = \sum_{m=0}^{\infty} \frac{(a)_m (b)_m}{(c)_m} \frac{z^m}{m!} \tag{A.73}$$

Again, this function is defined if $|z| < 1$. When $|z|$ is greater than 1, extrapolation is necessary. Figure A.37 shows the behavior of $_2F_1(a,b,c;z)$ for $a = 1.2\mathrm{i}$, $b = 1 + 4\mathrm{i}$ and $c = 1$, which are typical values used in the CDW approximation.

Appendix 11: Visibility of Interference Pattern

Distortion Factor and Interferences

To observe the oscillations in the angular distribution of emitted electrons, two conditions have to be fulfilled. Firstly, a term of the form $\cos f(\theta_d, d_{HH})$ or $Ae^{if(\theta_d,d_{HH})} + Be^{-if(\theta_d,d_{HH})}$ (θ_d is the detection angle and d_{HH} is the distance between the protons) has to appear in the expression of the autoionization amplitude. Secondly, the period has to be less than 90°.

In the Final State Interaction (FSI) approximation, the interference term originates from the introduction of the distortion factor $D_k(\overrightarrow{r_N})$. In the eikonal approximation, $D_k(\overrightarrow{r_N}) = \chi\, e^{-2i/k_P}$ with $\chi = k_P r_N - \overrightarrow{k_P} . \overrightarrow{r_N}$. Since k_P only depends on θ_d, this approximation cannot generate oscillations due to Young-type interferences. Consequently, it is necessary to use other approximations, such as Continuum Distorted Wave (CDW) approximation since $D_k(\overrightarrow{r_N})$ contains an exponential term that depends on both θ_d and d_{HH}.

Autoionization Amplitude in CDW Approximation

General Expression of the Distortion Factor

Following the theoretical work performed by Barrachina and Zitnik (cf. Chap. 5), the amplitude of probability is written

$$
\begin{aligned}
F \approx & -i\int_0^{\infty} t^{2i/k_P} e^{-(1-i\varepsilon)\Gamma t/2} dt \\
& -i\int_0^{\infty} \sum_{n=1}^{2} (D_n - 1)\exp\left[i(\overrightarrow{v} - v\overrightarrow{r}/r) . \overrightarrow{R_n}\right] t^{2i/k_P} e^{-(1-i\varepsilon)\Gamma t/2} dt
\end{aligned}
\tag{A.74}
$$

F. Frémont, *Young-Type Interferences with Electrons*, Springer Series on Atomic, Optical, and Plasma Physics 77, DOI 10.1007/978-3-642-38479-0,

Using CDW approximation, the distortion factor D_n can be expressed as

$$D_n = \Gamma(1 - 2i/v_n)\, e^{4\pi/v_n}\, {}_1F_1(2i/v_n, 1; i\chi_n) \tag{A.75}$$

where χ_n are the values of χ for each proton ($n = 1,2$).

In relation (A.74), only the second term, called F_I gives rise to the expected oscillations. When χ_n tends towards 0,

$$D_n \approx \Gamma(1 - 2i/v_n)\, e^{4\pi/v_n}\, e^{4i\chi_n/v_n}$$

Therefore

$$\sum_{n=1}^{2} (D_n - 1) \exp\left[i(\vec{v} - v\vec{r}/r) \,.\, \overrightarrow{R_n}\right] \approx (D_1 - 1)e^{i\vec{\chi}\,.\,\overrightarrow{r_H}} + (D_2 - 1)e^{-i\vec{\chi}\,.\,\overrightarrow{r_H}} \tag{A.76}$$

In relation (A.76), the second term leads to the oscillations.

Assuming $D_1 - 1 = D_2 - 1 = D$, the oscillating part of the autoionization amplitude is written $I(\theta_d, t) \approx D \cos g(\theta_d, t)$. Finally,

$$F \approx A - i \int_0^\infty g(\theta_d, t) h(\theta_d, t) dt \tag{A.77}$$

with

$$g(\theta_d, t) = \cos\left(\vec{\chi} \,.\, \overrightarrow{r_H}\right) \tag{A.78}$$

where $h(\theta_d, t)$ is a weighting function of the cosine function. The result for F_I is discussed for simple particular values of the function $h(\theta_d, t)$.

Visibility of the Interferences

$h(\theta_d, t) = \delta(t - t_m)$

Suppose that the function $h(\theta_d, t)$ is equal to 0, except for a value t_m of t. Thus, $F_I \approx \int_0^\infty \delta(t - t_m) \cos g(\theta_d, t)\, dt = \cos g(\theta_d, t_m)$

$$g(\theta_d, d_{HH}) = 2\left(v_N^2 t^2 + r_0^2\right)^{1/2}\left(1 + \frac{k_T^2}{k_P^2}\cos^2\theta_d\right) - k_T v_N t \cos\theta_d + v_N^2 t + r_0 k_T \sin\theta_d$$

Fig. A.38 Dependence of $g(\theta_d, t_m)$ with the detection angle

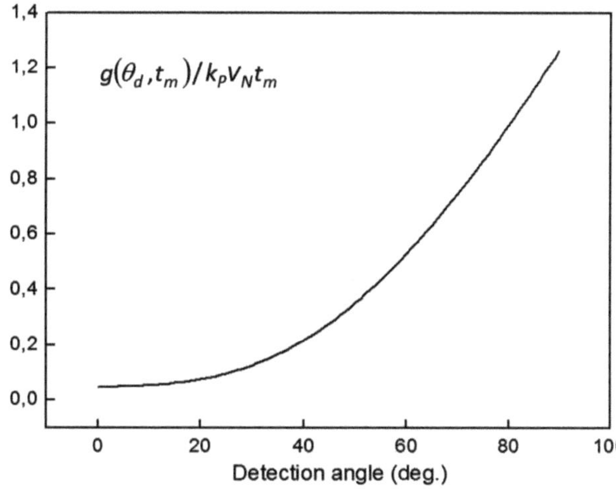

A typical value for t_m is that corresponding to the mean deexcitation time of the projectile, *i.e.* $t_m = 2/\Gamma \sim 400$ a.u. Therefore r_o can be neglected, and the above expression can be simplified.

$$g(\theta_d, d_{HH}) = v_N t_m \left[2\left(1 + \frac{k_T^2}{k_P^2}\cos^2\theta_d\right) - k_T \cos\theta_d + v_N \right] \tag{A.79}$$

The result for the expression in brackets is presented in Fig. A.38 as a function of the detection angle. This induces a smooth dependence in θ_d. Consequently, it can be replaced by a function of the type $a \cos\theta_d$. The term cos g is shown in Fig. A.39. Oscillations are revealed, with a pseudo-period of about 19°, which is consistent with the experiment.

$h(\theta_d, t) = e^{-\alpha t}$

In contrast with the previous approximation, we focus our attention on continuous dependence of $h(\theta_d, t)$ with time, which is more realistic than the peaking approximation. To simplify, it is assumed that $h(\theta_d, t)$ decreases exponentially with time, i.e. $h(\theta_d, t) = e^{-\alpha t}$, which is reasonable. To simplify the discussion, it is also assumed that the coefficient α is independent on the detection angle. Therefore, the amplitude that contains oscillations writes

$$F_I \propto \int_0^\infty e^{-\alpha t} \cos g(\theta_d, t)\, dt \tag{A.80}$$

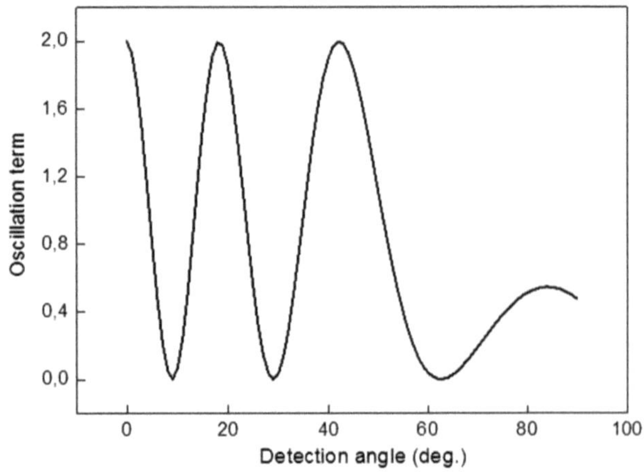

Fig. A.39 Oscillation term, from expression (A.79) for a typical deexcitation time of 400 a.u.

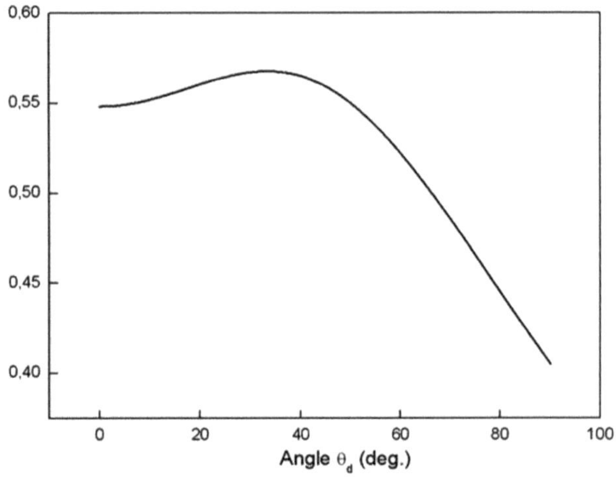

Fig. A.40 Interference term calculated using expression (A.80).

The integral is easily calculated and the result for the integral is shown in Fig. A.40. Oscillations are no longer visible, due to the fact that the cosine function is integrated from 0 to infinity.

Both extreme examples show that integration in time is the crucial problem in the possible visibility of the fringes. If time is concentrated in one point (peaking approximation), the theory is in agreement with experiment. In contrast, using the present autoionization theory, inclusion of infinite range in time destroys interference pattern.

Conclusion and Perspectives

Conclusion

Since the hypothesis of the wave-particle duality of light at the end of the seventeenth century, and the fascinating discovery of light interferences, this topic has never ceased to be the focus of experimental and theoretical studies. Despite the progress made, the nature of light, like the nature of matter, remains both a problem and a mystery. In some cases, light and matter can be described as punctual objects. However, in the presence of slits or atoms, the wave nature emerges and needs to be accounted for. Therefore, light and matter are no longer discrete, but rather extended objects in space and time. Consequently, when a single particle hits a slit or scatters on an atom, it is impossible to predict its position. Such a prediction can only be made for a large number of particles.

This book has addressed interferences using electrons in three ways:

- First, analogies with Young's works have been discussed.

  The ionization of molecules, in which one electron is ejected from two indistinguishable centers, is similar to the result obtained by Young in 1801. The amplitude of the waves induced by two sources add coherently, giving rise to series of maxima and minima in intensity.

  Experiments in which one electron passes through an electrostatic biprism or is ionized in the direction of a projectile, after double excitation of an atomic target, resemble the experiment Young performed in 1804 using a slip of card.

  Finally, only a few experiments are analogous to Young's double-slit experiment of 1807. These include Jönsson's experiment in 1961 with real slits, and the experiment performed by our group in 2005.
- Interferences have also been discussed with regard to the number of incident particles. Most of the experiments were conducted with photon or electron beams, and the number of particles was not controlled (Young 1807, Davisson and Germer 1927, Jönsson 1961). Then, experiments using strongly attenuated beams, in order to obtain single particle conditions on average, were performed with photons (Taylor 1909) and electrons (Merli 1976 and Tonomura 1989).

F. Frémont, *Young-Type Interferences with Electrons*, Springer Series on Atomic, Optical, and Plasma Physics 77, DOI 10.1007/978-3-642-38479-0,

In all of these cases, single impacts were observed on screens, and an interference pattern rebuilt after a long detection time. Finally, real single-particle conditions were created with photons (1985) and electrons (2005). The goal was, in the case of photons, to demonstrate their corpuscular behavior. In the case of electrons, the aim was to show that a single particle may behave like a wave. These two demonstrations can be considered as complementary.

- It is an acknowledged fact, since the middle of the twentieth century, that photon and electron interference experiments are analogous. Indeed, angular distributions show the emergence of series of maxima and minima in intensity whose period can be calculated using the same approximate formula. This analogy could be extended to other fundamental quantities, such as width distribution for example. All of the distributions are shown to oscillate periodically. However, in the case of photons, the angular distributions for the width and wavelength maximum do not oscillate continuously. This result is mainly due to the function that describes the angular distribution.

  Despite the analogy, unfortunately, strong differences remain in the theoretical approach. In contrast with photon experiments, for which calculations can be made analytically, no complete theory exists in the case of electron experiments following atomic collisions.

The study of interferences clearly shows both the richness of the discoveries made and the path that lies ahead. This is particularly true for collision-induced interferences, where the Coulomb force is involved in an N-body problem. At present, no solution exists to shed light on the oscillations caused by interferences, except when using a peaking approximation. This involves taking an average value for the Auger de-excitation time of the projectile that emits the electron, inducing average distances between the protons. Even if this approximation gives reasonable values for the period and for its evolution with the electron wavelength, the model cannot predict the electron energy distributions.

Given the limitations of the theory, some thoughts will be presented in the following section. In fact, three possibilities remain. The first is to replace the distortion factor by an exact one-electron two-center wave function. This method, which has been introduced in many cases, for example to treat electron scattering or photoionization, is useful for obtaining exact solutions for continuum states (see [6] and references therein). Secondly, since Auger de-excitation follows double electron capture, one can imagine introducing double-capture probability in the calculation. We will show that the separation between double capture and the Auger effect is no longer valid in some specific cases.

Finally, a semi-classical model has been developed very recently. Based on both the particle and wave behavior of the electron, this model is able to reproduce with good accuracy the oscillations observed experimentally. At present, there is no explanation for this success. However, one may suppose that the conventional view

[6] Hadinger, G. et al.: Continuum wavefunctions for one-electron two-centre molecular ions from the Killingbeck–Miller method. J. Phys. B **29**, 2951 (1996)

of interferences can be questioned depending on whether the model gives results that are entirely consistent with the experiment. Future work is thus needed in this direction, and new ideas on the nature of the particles are required.

Theoretical and Experimental Perspectives

Exact Solution of the Continuum Wave Function

In Final State Interaction (FSI) approximation, the amplitude can be given as

$$A \propto \int_0^\infty \psi\left(\overrightarrow{r}, \overrightarrow{k_T}\right) t^{\frac{2i}{v_N}} e^{-(1-i\varepsilon)\frac{\Gamma t}{2}} dt \tag{C.1}$$

To build the $\psi\left(\overrightarrow{r}, \overrightarrow{k_T}\right)$ wave function associated with the electron emitted in the Coulomb field of the two protons, spheroidal coordinates (ξ, η, φ) are used

$$\begin{aligned} \xi &= \frac{r_1 + r_2}{R} \\ \eta &= \frac{r_1 - r_2}{R} \end{aligned} \tag{C.2}$$

In these expressions, r_1 and r_2 are the distances between the electron and each proton, and R is the distance between the two protons. The azimuthal angle φ is defined in the 0–2π range.

The Schrödinger equation describing the time evolution of the electron gives rise to a wave function that can be written as an infinite sum of partial wave functions

$$\psi\left(\overrightarrow{r}, \overrightarrow{k_T}\right) = \frac{1}{\sqrt{k_T}} \sum_m \sum_\ell Y^*_{\ell m}(k_T, \cos\theta, \varphi)\, Y_{\ell m}(k_T)\, J_{\ell m}(k_T)\, e^{-i\delta_{\ell m}} \tag{C.3}$$

In this expression, $Y^*_{\ell m}$ and $Y_{\ell m}$ are spherical harmonics used for the quasi-radial part, and $J_{\ell m}$, which decomposes into a sum of Bessel functions, characterizes the radial part. The ℓ and m quantities are, respectively, the well-known angular momentum and magnetic quantum number. The coefficient $\delta_{\ell m}$ is a phase term.

The amplitude is obtained by replacing expression (C.3) into (C.1). In principal, one has to take into account all the possible ℓ and m values. Nevertheless, in order to get some first indications of a possible result, certain crude approximations can be made. First, we make the molecular axis perpendicular to the beam direction. Therefore, $\eta = 0$ and $\xi = 2r/R$. Then, since the electron velocity is of the order of 1 a.u. at angles much smaller than 90°, only small values of ℓ can be introduced. Thus, retaining only $\ell = 0$, the integral can be written

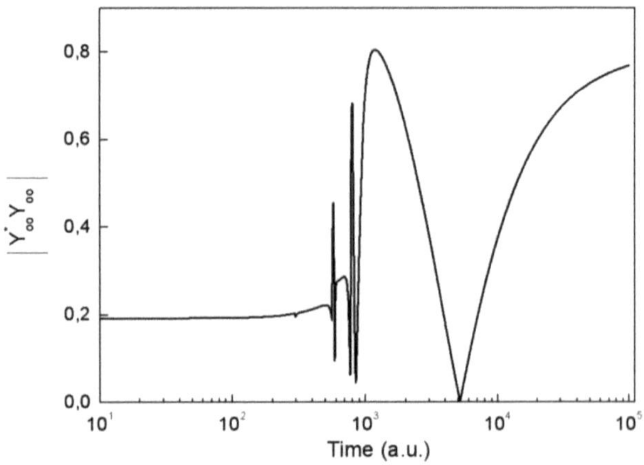

Fig. C.1 Contribution of $\left|Y^*_{oo}\,Y_{oo}\right|$ as a function of time, for an electron energy of 33 eV

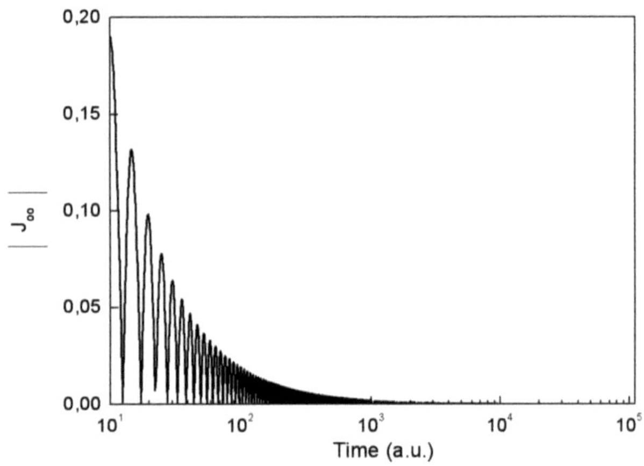

Fig. C.2 Contribution of $|J_{oo}|$ as a function of time, for an electron energy of 33 eV

$$A \propto \int_0^\infty Y^*_{oo}\,Y_{oo}\,J_{oo}\,e^{-i\delta_{oo}}\,t^{\frac{2i}{v_N}}\,e^{-(1-i\varepsilon)\frac{\Gamma t}{2}}dt \tag{C.4}$$

To understand the role of the angular and radial contributions, the quantities $\left|Y^*_{oo}\,Y_{oo}\right|$ and $|J_{oo}|$ are represented as a function of time (Figs. C.1 and C.2) at an electron energy of 33 eV, since the dependence on energy is weak.

The radial part is constant in the 0–600 a.u. range, then reaches maximum at ~1,500 a.u. and at infinity. In contrast, $|J_{oo}|$ decreases with increasing time, and strongly oscillates. The product of both $\left|Y^*_{oo}\,Y_{oo}\right|$ and $|J_{oo}|$ quantities, represented in Fig. C.3, has approximately the same behavior as that of $|J_{oo}|$.

Therefore, this product can be fitted using the following formula

$$\left|Y^*_{oo}\,Y_{oo}\,J_{oo}\right| \approx a\;e^{-\gamma t}\sin\omega t \tag{C.5}$$

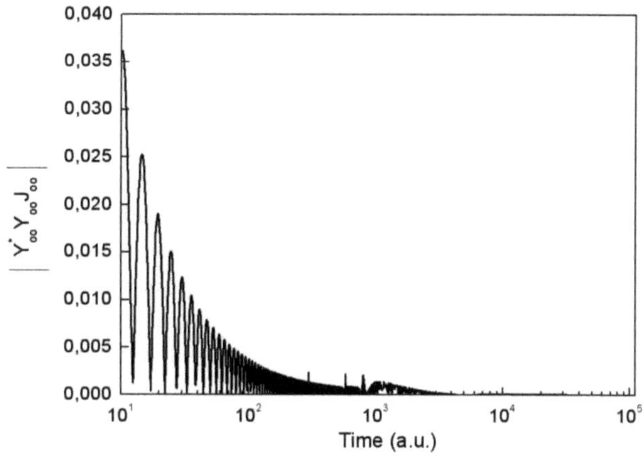

Fig. C.3 Contribution of $\left|Y^*_{oo}\, Y_{oo}\, J_{oo}\right|$ as a function of time, for an electron of energy 33 eV

and the amplitude is finally

$$A \propto \int_0^\infty e^{-\gamma t} \sin \omega t \; e^{-i\delta_{oo}} \; t^{\frac{2i}{v_N}} \; e^{-(1-i\varepsilon)\frac{\Gamma t}{2}} dt \tag{C.6}$$

Unfortunately, the integral limits are not reduced, so that the oscillations are not expected to be revealed. This preliminary result shows that zero order is not valid. One would expect higher ℓ and m values to play an important role.

Introduction of the Double-Capture Process

The Role of Double Capture at Large Projectile Velocities

As discussed in Chap. 5 on the subject of autoionization theory, double capture and autoionization have been separated. The argument for this approximation lies in the characteristic times of each process. While double capture time is of the order of a few a.u., the Auger process takes about 100 times longer at a projectile velocity of 0.63 a.u. However, when looking at the specific problem of interferences, one has to bear in mind that these are due to the distortion of the electron wave function owing to the presence of the two protons. Throughout the discussion, the distortion was assumed to start at the time of autoionization. This assumption is in fact wrong, since the electron wave function is already distorted during the capture process. In this sense, even at relatively high projectile velocities, the separation of collision and post-collision processes is not valid. In other words, it seems that, to correctly interpret the experimental oscillations, the capture process has to be included in the theory. The capture process and the Auger effect can even be considered as independent.

The Role of Double Capture at Very Low Projectile Velocities

A Five-Body Problem

When the projectile velocity is considerably reduced, the characteristic times for capture and autoionization become of the same order of magnitude. For example, at a projectile velocity of 0.06 a.u., corresponding to an energy of 300 eV that can be achieved with ECR sources, both times are about 200 a.u. Therefore, the two processes are mixed, and the electron emission occurs in the field of three centers, instead of one. In this case, one can easily understand that the separation of the processes is no longer valid. Each is dependent on the other. Theoretically speaking, this means that the problem to solve is a true five-body problem, including two protons, two electrons and the projectile.

Visibility of Interferences

At low projectile velocities, the electron is emitted in any direction from a molecule consisting of three centers. Due to the fact that the molecule is randomly oriented, there is no preferred emission direction for the electron. Therefore, interferences are expected to be negligible.

Nevertheless, another scenario should be taken into consideration. Several years ago, double capture in $O^{5+} + H_2$ and $N^{7+} + H_2$ collisions were studied, experimentally and theoretically, at a projectile energy of 100 eV [7, 8]. After double capture, the protons separated with a velocity v_H ~0.0195 a.u. at infinity when there was no influence of the projectile on the protons. Thus, at relatively high projectile velocities, the fragmentation is found to be isotropic, and the energy distribution is given by a sharp peak centered at ~9.5 eV (top of Fig. C.4), which corresponds to a free fragmentation.

For impact velocities ranging from ~0.1 to 0.5 a.u., the fragments are found to be emitted in the backward direction with respect to the incident beam direction, i.e., in the opposite direction of the projectile, suggesting the emergence of the role of Coulomb forces induced by the projectile field. Moreover, the fragment energies are lower than 9.5 eV (middle of Fig. C.4). At very low impact velocities (bottom of Fig. C.4), highly energetic fragments are observed at forward angles, i.e., in the direction of the projectile. This surprising effect was understood by theoretically investigating energy distributions as a function of the impact parameter. While at velocities of the order of 0.5 a.u., double capture predominantly occurs at large impact parameters (soft collisions), the contribution of small impact parameters

---

[7] Sobocinski, P. et al.: Evidence for highly energetic fragments following electron capture in O^{5+} + H_2 collisions at low impact velocities. J. Phys. B **35**, 1353 (2002)

[8] Sobocinski, P. et al.: Energy distributions of protons following electron capture in collisions of N^{7+} ions with H_2 at sub keV impact energies. In: AIP Conference Proceedings **680**, 40 (2003)

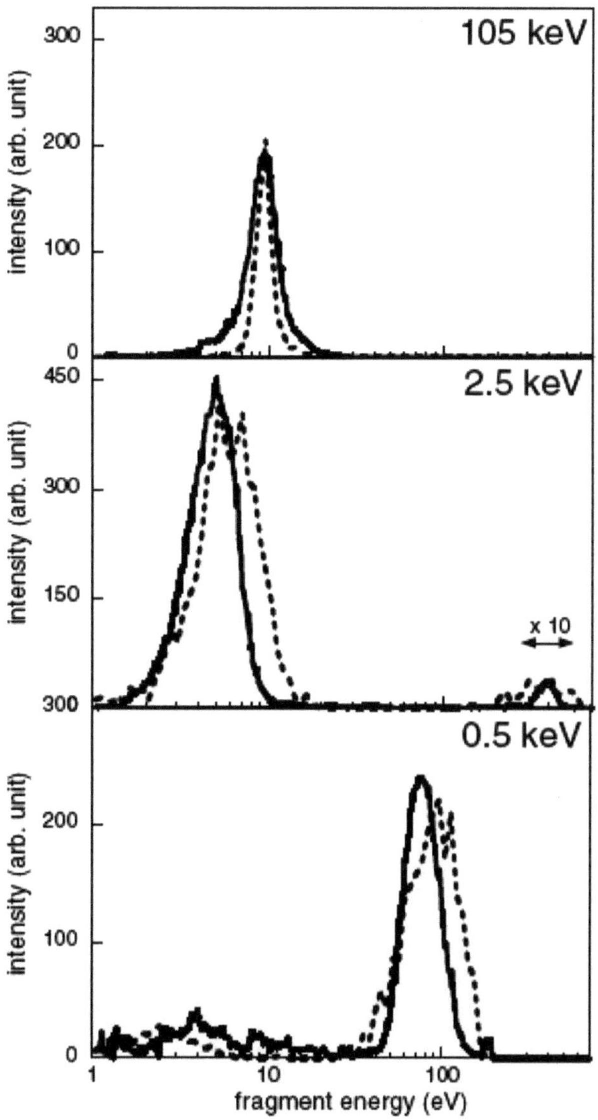

Fig. C.4 Fragment energy distributions following $O^{5+} + H_2$ at three projectile energies. *Dashed curves*, experiment; *full curves*, result of quasi classical calculations

(hard collisions) increases to become dominant at projectile velocities lower than 0.1 a.u. Hence, soft collisions are responsible for the ejection of protons in the backward direction, and hard collisions give rise to protons that are emitted at forward angles.

This main result is summarized in Fig. C.5 for the collision system $N^{7+} + H_2$ at projectile energies of 1.4 keV (top of the figure) and 32 eV (bottom of the figure). At the highest energy, the dominant process is the emission at backward angles, since

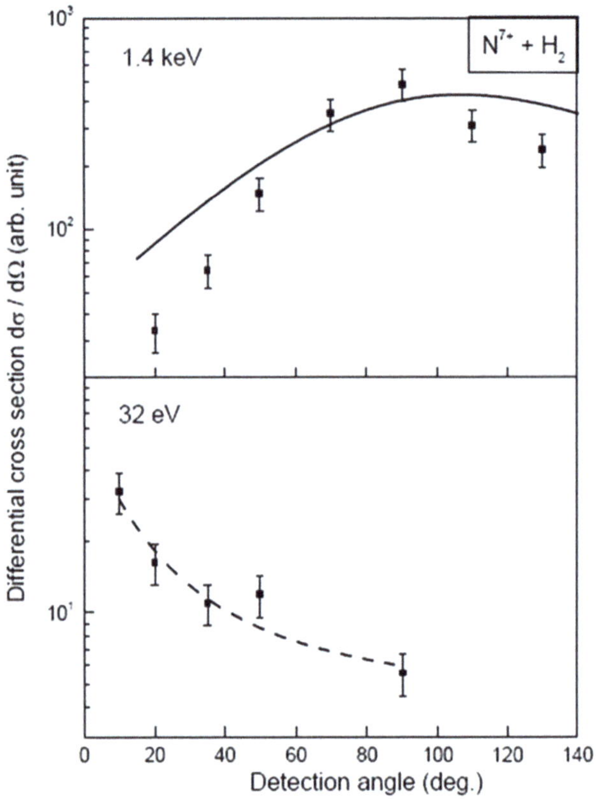

Fig. C.5 Cross sections for detecting one proton as a function of the detection angle, for the system $N^{7+} + H_2$ at projectile energies of 1.4 and 32 eV. The *solid line* is the result of the quasi classical calculation, and the *dashed line* is a guide for the eye

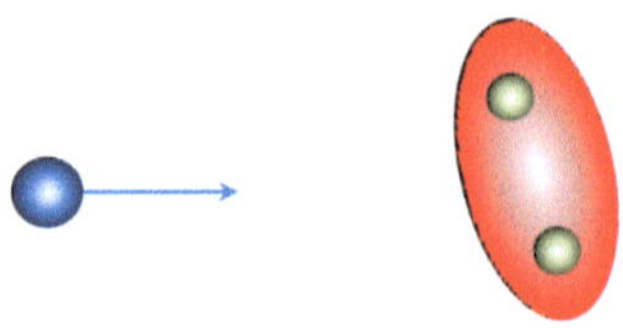

Fig. C.6 The He projectile approaches the molecular target at small impact parameters

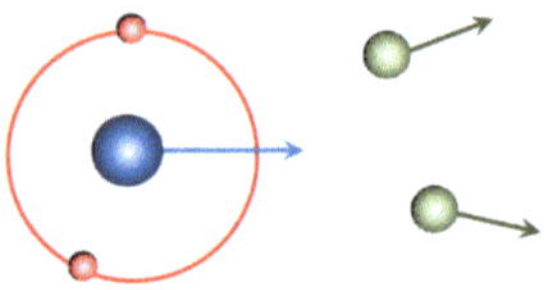

Fig. C.7 The projectile captures the electrons from the target and pushes the protons in the forward direction

the protons are repulsed by the projectile after double capture. In contrast, at the lowest projectile energy, protons of energies of the order of 30 eV are pushed in a forward direction, indicating, firstly, collisions at very small impact parameters and, secondly, that double capture occurs before the projectile reaches the target.

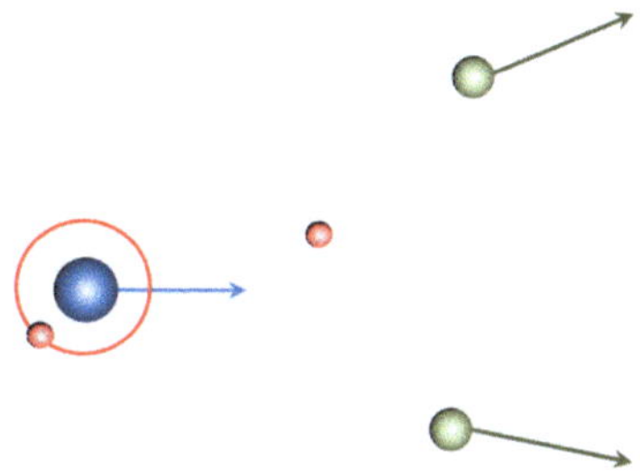

Fig. C.8 The projectile emits a Auger electron, which interferes on the protons at forward angles

This result is decisive with regard to the study of interferences. Given what has been said previously, the following scenario can be considered (Fig. C.6). To capture both H_2 electrons at very low impact velocity, the He projectile approaches the target at small impact parameters (Fig. C.6). During this process, the protons are emitted at forward angles (Fig. C.7), and the He projectile emits an Auger electron (Fig. C.8). Thus, to produce interferences, the electron has to scatter at forward angles, contrary to what happens at higher projectile velocities. The study of the evolution of interferences in the angular distribution with the projectile velocity is expected to provide information on the capture process itself.

Index

A
Angular distributions, 212
Asymmetric collisions
  defined, rainbow effect, 55
  doubly differential cross section, 56, 57
  electron captured, 'way-out' of collision, 54, 55
  first transition, entrance and outgoing channels, 54
  O^{2+} + He collision, single electron capture, 55
  potential curves, 53, 54
  scattering angle dependence, impact parameter, 56
  single capture differential cross section, 56–57
  Stueckelberg oscillations, 55
Atomic collisions
  approximations
    electronic basis, 46
    projectile motion, 45–47
    separation, 45
  collisional process
    description, 42
    excitation, 42, 43
    ionization, 42, 43
    principal quantum number n and angular momentum ℓ, 42, 43
  electron emissions (*see* Electron emissions, atomic collisions)
  post-collisional process
    emission, Auger electron e_A, 44
    He^{2+} ion and molecular target H_2, 42
    radiative cascade, two electrons, 42, 44
Atomic units
  electron, circular motion, 152–153
  electron energy, 153
  international system, 151–152
  quantities, 153
Atoms interferences
  double-slit experiment, 23–24
  Ramsey-Bordé interferometry, 26–28
  three-grating experiment, 24–26
Attenuated electron beam interferences
  analogous experimental set-up, 39
  electron interference fringe pattern, 38, 40
  electrostatic and Fresnal biprisms, 37, 39
  fringes, 38
  Italian group's experiments, 39–40
Auger de-excitation time, 212
Auger electron emission, Young-type interferences
  ion-atom collisions, 128–137
  ion-molecule collisions, 137–140
  quantum mechanics, 125–128
  semi-classical approach, 140–149
Auger lines interferences
  electron energy spectra, 1.4 keV He^+ + He collisions, 96–97
  spectrum calculation, 97
Autoionization (AI) interferences. *See* Direct ionization (DI) interferences

B
Barker and Berry model
  Auger electrons, 166
  collision partners, 166

F. Frémont, *Young-Type Interferences with Electrons*, Springer Series on Atomic, Optical, and Plasma Physics 77, DOI 10.1007/978-3-642-38479-0,

Barker and Berry model (*cont.*)
features, PCI, 167–168
BE. *See* Bose-Einstein (BE) condensates
Bose-Einstein (BE) condensates
interference pattern, 32, 33
laser and evaporation cooling, 32
magneto-optical trap, 32, 33

C
CDW. *See* Continuum distorted wave (CDW)
Collision diagram, 199, 200
Collision-induced interferences, 212
Continuum distorted wave (CDW)
angular distribution, emitted electrons, 137
approximation, 207–208
contour plot, 2s2 1S, 137, 138
detection angles, spectra calculation, 136–137
distortion factor, 136
electron and projectile, Coulomb interaction, 135
Continuum wave function
azimuthal angle φ, 213
FSI, approximation, 213
$|J_{00}|$ contribution, 214
Schrödinger equation, 213
spherical harmonics, 213
$|Y^+_{00}\ Y_{00}|$, 214
$|Y^+_{00}\ Y_{00}\ J_{00}|$ contribution, 214, 215
Corpuscular vision
diffraction, 4
photoelectric effect, 9
reflection and refraction, 4
separation, light, 4

D
De Broglie wave-particle duality
BE condensates, 23, 24
electron beam, crystal and collector, 20, 21
electron diffraction, 20
intensity, elastic scattering, 20, 21
maxima and minima intensity, 23
monochromatic beam, neutrons, 22
neutron interferometer, 20, 22
non-monochromatic thermal neutrons, 21
quanta, 19
resonant level, 23
scattered and transmitted beams, 23, 24
spatial-temporal model, 19
thermal neutrons, 23, 24
wave theory, 19
DI. *See* Direct ionization (DI) interferences
Direct ionization (DI) interferences
Auger electron emission, 2.4 keV $e^- + D_2$ collisions, 95, 97
Auger profiles
H^+ + He collisions, 95, 96
36 keV H^+ + He collisions, 94, 95
collision system, 94
destructive/constructive, 96
Distortion factor and interferences
angular distribution, emitted electrons, 207
CDW, 207–208
FSI, 207
Double-capture process
large projectile velocities, 215
small projectile velocities (*see* Small projectile velocities)
Double-slit experiment
bright and dark fringes, 23
detector plane, 23, 25
scanning-electron microscope, 23, 25
supersonic jet, 23

E
Elastic collisions
antibonding orbitals, 49, 50
bonding orbitals, 49
ECR sources, 46
oscillations, 48
perturbations, distorted potentials, 48
potential energies, He^+ + He collision system, 49, 50
projectile deflection and trajectories invertion, 49, 51
scattering differential cross sections, He^+ + Ne and He^+ + He, 46–48
trajectories, same scattering angle, 49, 50
Electron emissions, atomic collisions
Young 1801
angular distribution, 61, 63
contribution, different ℓ values, 59, 60
electron capture, 83–84
emitted electron velocity, 60, 61
energy hν ionization, 58, 59
experimental and theoretical ratios, 62, 64
experimental set-up, N_2, 61, 62
free-field spherical waves, 59
H_2 ionizations (*see* H_2 ionizations)
interference term, $1 + \sin kR / kR$ plot, 60
ionization interferences, 57
localization and de-localization, 61

N_2 and O_2, photon energy, 58, 59
quantum numbers, 58
reduced cross sections, hydrogenic atoms, 62, 63
Young 1804
Auger electron, 92
Coulomb path interference effect, 91
spectra interpretation, 92–94
Electron energy distribution
calculation ratio, 186–187
detection angle, 188–189
energy maxima, 183–184
initial intensity, 189
oscillation amplitude, 183
post-collision effect, 184
protons, 184
sine function, 185–186
Electron interferences
macroscopic interferometers (*see* Macroscopic interferometers)
nanoscopic interferometers (*see* Nanoscopic interferometers)
revisited
E_{max} angular distribution, 120–121
energy distribution, fixed angle, 121–122
Electrons, ion interferences
elastic collisions, asymmetric and symmetric (*see* Elastic collisions)
electron capture
asymmetric collisions (*see* Asymmetric collisions)
symmetric collisions (*see* Symmetric collisions)
Electron spectra, FFT
120° and 160°, Auger spectra, 197
Auger spectrum, 196
Energy and momentum conservation laws, 199–200

F
Fast Fourier Transform (FFT)
application, electron spectra (*see* Electron spectra, FFT)
decomposition, periodic function, 191
Gaussian curves, 191, 192, 197, 198
non-periodic function (*see* Non-periodic function, FFT)
pseudo-period, oscillations, 121–122
sampled function, 193–196
Fermi's golden rule, 158–159
Feynman thought experiment
bullets, toward two holes, 37, 38
description, 37
detectors, electrons, 37
interference fringes, 37
quantum mechanics, 36
source, electrons scatter through two holes, 37, 38
FFT. *See* Fast Fourier Transform (FFT)
Final state interaction (FSI) approximation, 207
First order interferences
Bessel function, 66
Born calculations, 66
cross-section ratios, detection angles, 66
DDCS measurement, 68
detection, 2 eV up to 200 eV, 64
experimental to theoretical ratios, 66, 67
fully measured DDCS ratio, 68, 69
photon energy, 64
spectra realized, 30° and 150°, 65
FSI. *See* Final state interaction (FSI) approximation
Full width at half maximum (FWHM), 163
FWHM. *See* Full width at half maximum (FWHM)

G
Gamma functions, 203–204

H
H_2 ionizations
Born approximation, 159–160
electron impact
angular distribution, emitted electrons, 88, 89
electron trajectories, 90, 91
experimental and theoretical, doubly differential cross section ratios, 87, 88
experimental ratio, $\sigma(H_2)/\sigma(He)$, 90, 91
function, detection angle, 85, 86
ion/electron beams, 85
2.4 keV $e^- + D_2$ collisions, 87
molecular and atom, TDCS, 85–86
multiply charged ions, 85
projectiles, 85
TDCS, ionization of He and H_2, 88–90
Fermi's golden rule, 158–159
ion impact
first order interferences, 64–69
frequency parameter, 81–82

multiple-order effects (*see* Multiple-order effects, H_2 ionizations)
multiple order interferences, 69–73
projectile nucleus, interferences, 82–83
ionization amplitude, probability, 157–158
probability amplitude, 160–161
Hypergeometric functions
definition, 204–205
function ${}_1F_1$, 205–206
function ${}_2F_1$, 206

I

Interferences
incident particles, 211
massive particles (*see* Massive particles interferences)
Ion-atom collisions, Auger electron emission
capture parameters, 129
CDW approximation, 135–137
DI and AI, 129, 130
eikonal approximation, wave function
contour plot, 134, 135
dashed and full curve, electron energy, 133
3D plot, 2s2 1S, 134, 135
electron energy profile, 129, 131
He configurations, 129
impact parameter distributions, 130, 131
influence, time t_o, 130, 131
Ionization of molecules, 211
Ion-molecule collisions
Auger electron emission
contour plot, angular distribution, 139, 140
expression, transition amplitude, 137–139
oscillations, 139, 141
electron, 172
molecular orientations, 173–174

J

Jönsson's interference experiment, 35, 36, 211

K

Kirchhoff diffraction theory, 30, 31

L

Light interferences, attenuated beam, 9–10

M

Mach-Zehnder interferometer, 26
Macroscopic interferometers
attenuated electron beam (*see* Attenuated electron beam interferences)
Feynman thought experiment, 36–38
Young experiment, electron beam, 35–36
Massive particles interferences
atoms, 23–26
BE condensates, 32–33
De Broglie wave-particle duality (*see* De Broglie wave-particle duality)
grating interferometer, 24, 25
intensity, 26
large molecules and clusters
diffraction pattern, 30
fringes, 32
fullerenes, 30
interference patterns, 30, 31
Kirchhoff diffraction theory, 30, 31
QMS, 32
quantum interference, 32
wave nature, 30
neutrons, 20–23
small molecules
atomic sources, 28, 29
dimers, 28, 29
interferometry, 28
Multiple-order effects, H_2 ionizations
cross-section ratio calculation, 77
CTMC and CDW, 74
direct ionization and emission, 80
electron spectra, 79
experimental and theoretical ratios, 74, 75
experimental-to-theoretical and theoretical ratios, 76
He^+ + He, He^{2+} + He and e^- + He, 80
interference pattern, 77, 78
20 keV H^+ + H collisions, 78
polynomial function, log-log scale, 79
second-order interference effects, 75
TDCS, 73–74
Multiple order interferences
Bessel function, 71
coherent emission, 73
energies of 1–5 MeV, experimental ratios, 71, 72
experimental to theoretical ratios, three angles, 70
fitted curves, 72, 73
ion-atom and ion-molecule collisions, 69
waves, emitted electron scatters, 69

N

Nanoscopic interferometers
- Coulomb-type interaction, 41
- electron source, 41
- interference study, 41
- process, atomic collisions (*see* Atomic collisions)

Neutrons interferences, 20–23
Non-periodic function, FFT
- definition, 192
- dirac delta distribution, 193
- rectangular distribution, 193, 194
- sine function, 193

O

One and two event probabilities
- Auger effect, 106
- collision parameters and single electron condition, 107–108
- experimental device, 108–109
- independent collisions, 106–107
- interference condition, one electron, 106–107
- low energy ion source, 108

One-electron two-center wave function, 212

P

Path interference model
- ion-atom collisions, 171–172
- ion-molecule collisions, 172–174
- parameters, 174
- young-type double slit interferences, 174–176

PCI. *See* Post collision interaction (PCI)
Photomultipliers (PMs), 11, 12
Photon and electron experiments, 212
Photon interferences
- ancient Greek civilization, 1
- Arab contribution, 2
- coherence, time and space, 13
- conceptions, light, 1
- contrast and visibility
  - definition, 13–14
  - electron emission, 14
  - electron spectra, 14
  - maximum and minimum intensity, 14
  - polychromatic source, 15–16
  - spatially extended source, 16–17
- corpuscular vision, light, 4
- discovery and formulation
  - double-slit experiments, 6, 7
  - electron interference experiments, 6
  - tides, 6
  - waves mixture, 5
- displacement, fringes, 7, 8
- electromagnetic waves, 7
- experiment
  - analogy relevance, 117–118
  - young double-slit experiment, 118–120
- light interferences, 9–10
- Michelson interferometer, 7, 8
- occident light
  - light ray, 3
  - optics studies, 2
  - reflection and refraction laws, 3
  - telescope, 3
- quanta revolution and corpuscular vision, 8–9
- single photons
  - Mach-Zender type interferometer, 11
  - PMs, 11, 12
  - Poisson probability law, 10
  - Taylor's experiment, 10
- total intensity, 12
- visualization, undulations, 3
- wave packets, 13
- wave vision, 4–5

PMs. *See* Photomultipliers (PMs)
Pochhammer coefficients, 204
Polychromatic source
- interference patterns, 15
- monochromatic source, 15
- total intensity, 15

Post collision interaction (PCI)
- angular dependence, 168–169
- Auger energy spectra, 163–164
- Barker and Berry model, 166–168
- electron, 166
- energy distributions, 167
- FWHM, 163
- low energies and high energy, 165–166
- principle, 163

Punctual objects, 211

Q

QMS. *See* Quadrupole mass spectrometer (QMS)
Quadrupole mass spectrometer (QMS), 32
Quantum mechanics, Auger electron emission
- amplitude, transition probability, 128
- autoionization process, 126–127
- collisions He^{2+} + He and He^{2+} + H_2, 125–126
- distortion factor, 128

Quantum mechanics (*cont.*)
emitted electron and projectile, interaction, 128
probability amplitudes, 127
wavefunction, 127

R
Ramsey-Bordé interferometry
interference pattern, 27
interferometers, 27
Mach-Zehnder interferometer, 26
momentum conservation, 26, 27
photon, energy, 26, 27
Recoil momentum
parallel component, 200–201
perpendicular component, 201–202

S
Scattering cross section
definition, 155
impact parameter function, 155–156
particle trajectory, 155, 156
Semi-classical approach, Auger electron emission
angular distribution, He^{2+} + H_2 collisions, 148–149
description, 140–141
droplets and trajectories, 142–143
protons and electrons, corpuscular aspect
forces, 143, 144
initial conditions, 144–145
trajectories, 145–147
walker, 141
wave behavior, 145, 147–148
Single-electron experiment
analogy, photon, 101
Bessel function, 104–105
coherence time and length, 106
contrast limitations, 104–105
energy and angular distribution, 105, 106
fixed-in-space molecules, 104
ion beam collides, 103
nanoscale interferometer, 101–102
photon interferences, 103
Small projectile velocities
five-body problem, 216
visibility, interferences
Auger electron, 219
captures, electrons, 218, 219
fragment energy distributions, 216, 217
He projectile approach, 219
O^{5+} + H_2 and N^{7+} + H_2 collisions, 216
proton detection, detection angle, 217, 218
Symmetric collisions
experimental observations
low/fast charged ions and atoms, 49
projectile energy, probability, 51
reciprocal velocity $1/v_p$, probability, 51, 52
interpretation, 52–53

T
TDCS. *See* Triple differential cross section (TDCS)
Triple differential cross section (TDCS)
electron emission, 73
ionization, He and H_2, 88, 90
molecular and atoms ratios, 86

V
Visibility of interferences
$h(\theta_{d,}\ t) = \delta(t-t_m)$
$(\theta_{d,}\ t_m)$ dependence, detection angle, 209
description, 209
$h(\theta_{d,}\ t) = e^{-\alpha t}$
autoionization theory, 210
description, 208–209
interference term calculation, 210
oscillation term, expression, 209, 210

W
Wavelength distribution
expression, 177–178
intensity and FWHM oscillations, 179, 180
λ *max*, 180–181
Wave visions
reflection and refraction, 5
spherical wavefront, 5
Width distribution, 177–178

X
X-ray diffraction, molecules, 112

Y
Young double-slit experiment
experimental set-up, 118
photon and electron interference, 120–121
photon wavelength, 122

wavelength distribution spectrum, 119
Young experiment, electron beam
coherent conditions, 35
interference pattern, after electrons scattering, 36
Jönsson's interference, slits production, 35, 36
principle, 35
Young-type double slit interferences
oscillation period, 175, 176
oscillations amplitude, 175, 176
photons, 174
Young-type electron interferences
electron interferences, oscillations
amplitude, 114
Auger spectra, 115
Bessel function, 116–117
FWHM, 115
FFT, 123
one and two event probabilities (*see* One and two event probabilities)
oscillations evolution, interference parameters, 112–114
photon interference experiment, 117–122
single-electron experiment (*see* Single-electron experiment)
spectral analysis
angular distribution, 112
fitting procedure, 111–112
30keV He^{2+} + H_2 collisions, 109–110
oscillation amplitude, 111
pseudo period, 110

Z
Zero order curve, Bessel function, 112, 113

MIX
Papier aus verantwortungsvollen Quellen
Paper from responsible sources
FSC® C105338

If you have any concerns about our products,
you can contact us on
ProductSafety@springernature.com

In case Publisher is established outside the EU,
the EU authorized representative is:
Springer Nature Customer Service Center GmbH
Europaplatz 3, 69115 Heidelberg, Germany

Printed by Libri Plureos GmbH
in Hamburg, Germany